PRIMES AND PROGR

PRIMES AND PROGRAMMING
An Introduction to Number Theory with Computing

Peter Giblin

University of Liverpool

CAMBRIDGE
UNIVERSITY PRESS

PUBLISHED BY THE PRESS SYNDICATE OF THE UNIVERSITY OF CAMBRIDGE
The Pitt Building, Trumpington Street, Cambridge CB2 1RP, United Kingdom

CAMBRIDGE UNIVERSITY PRESS
The Edinburgh Building, Cambridge CB2 2RU, United Kingdom
40 West 20th Street, New York, NY 10011–4211, USA
10 Stamford Road, Oakleigh, Melbourne 3166, Australia

First published 1993
Reprinted 1994, 1997

Printed in the United Kingdom at the University Press, Cambridge

A catalogue record for this book is available from the British Library

ISBN 0 521 40182 8 hardback
ISBN 0 521 40988 8 paperback

Contents

Preface

Whole numbers are a very basic part of everyone's mathematical experience. The fact that some numbers, like 17, do not break up into factors, whereas others, like 15, do, must surely have intrigued and attracted most people at some time, probably an early one, in their education. Number theory has turned out to have very practical applications in recent years, and this very basic distinction, between primeness and compositeness, has acquired practical importance on top of its inherent fascination. The numbers in question tend to be very much bigger than 15 and 17, of course, but then the fascination is greater, not less.

This book is about the theory of whole numbers and its underlying theme is the pair of questions: how can we prove a number is prime, and how can we factorise a number which is not prime? Here are some related questions: how can we list primes efficiently, either all the primes starting with 2 and going up to some limit, or just the primes between, say 1 000 000 and 1 000 100? How can we tell quickly, and with a high degree of confidence (but maybe not actual proof), that a large number is prime? Many methods have been devised over the past century or so for answering these questions, and in this book I shall describe some of them. I have confined myself to methods which use only 'elementary' number theory, that is which do not demand a knowledge of analysis or the kind of algebra which leads to the study of number fields. Everything here is developed from scratch, so that there are no particular mathematical prerequisites. Although some of the material, particularly that on orders and primitive roots, has a group-theoretic interpretation, I have chosen not to emphasise this.

Many standard techniques and results of number theory have turned out to play an important role in attacking the above questions, for example the venerable theories of continued fractions and quadratic residues. Although I have certainly strayed from the straight and narrow path of primality and factorization, the development of a topic such as continued fractions is strongly influenced by what is needed for the understanding of the continued fraction method of factorization. Nevertheless the reader will find a reasonably full account of the beginnings of this theory (and of quadratic residues, reciprocity, etc.) in the book.

With the advent of computers the domain of numbers which can be handled in practice has increased enormously, and it is an essential feature of this book that computers are used throughout to put the theory into practice. There is a choice to be made here: should one make use of a ready-made number theoretic package, which performs arithmetic operations such as multiplication to any number of figures (so-called multiprecision arithmetic), or should one write the programs oneself and be content with the size of numbers which ordinary languages such as PASCAL can handle? I have opted for the latter alternative, since I believe emphatically that turning mathematics into algorithms and programs is a worthwhile task which enhances one's understanding of the mathematics. (It is also true in my experience that students take more readily to programming when it solves questions which are concrete and interesting in themselves.) Thus the full excitement of handling huge numbers of 50 or 100 digits is denied to us, but we can perfectly well get the flavour of the subject through calculations with numbers of more modest size, say up to 16 digits.

I have borne in mind that there is a substantial connection between elementary number theory and 'recreational mathematics' , through strange or curious properties of whole numbers (there are even numbers called 'weird numbers' !). Some of the projects and computer exercises exploit this connection. After all, the only contact that many people maintain with mathematics after finishing their formal studies in the subject is a contact with recreational mathematics (and that is much better than no contact at all).

I have chosen PASCAL as the language of the book since it is fairly widely known, and is a structured language which can be read easily and if necessary converted into other similar languages. I used Turbo PASCAL but none of the special features of this language is exploited (except to a limited extent in Chapter 7, where there is some string-handling) and the programs should compile and run with virtually any implementation of PASCAL. Nearly all the programs needed are listed in full, and there are no very sophisticated progamming techniques used. In fact in this as in other branches of mathematics it is surprising how much of real significance can be achieved with a minimum of programming expertise.

The material of the book, with some of the theory and proofs omitted, was used at Liverpool University in a final (third) year course, but the material is equally suited to presentation at the second year level, and substantial parts of it are sufficiently elementary to be used earlier than that. A feature of the course was project work and a great many projects are included here. These are more substantial than mere 'computing exercises' and in some cases may entail looking something up from elsewhere. I am grateful to the many students who took my course over a period of four years for helping to remove ambiguities and other unintended difficulties from the projects. I am always impressed by the amount of trouble a student will take over a piece of work which is unique to him or herself. Naturally, the theoretical and computing exercises here have been through the same polishing process and they, too, have emerged brighter and cleaner than before. Any remaining mistakes are, of course, my responsibility and not the students'.

Much of the material for projects has been gleaned from the pages of such invaluable periodicals as *American Mathematical Monthly, Mathematics Magazine* and *Mathematics of Computation*, and I recommend these and similar periodicals as an endless mine of good ideas. I have quoted sources in many cases, and references (to the list at the end of the book) are always of the form [Bloggs (1987)], with square brackets. A statement such as (Euler (1760)) is merely a piece of information that Euler proved this result in 1760. It is of interest, it seems to me, whether a result dates from 1760, 1860 or 1960, and I have tried to indicate the dates of the more significant developments. However this is not a historical work and the reader must turn to the compendious pages of books such as [Dickson (1966)] for more information on this topic. Dickson's book was first published in 1919 so naturally it does not cover very much of the twentieth century history. There is quite a lot of information on the history of number theory in [Kline (1972)].

A few items are labelled with an asterisk * and these can be omitted without disturbing the logical flow of the book.

Besides the students who have contributed strongly to the final form of this book, I thank most warmly my colleague Professor J.W. Bruce, who contributed some substantial problems and projects, and also a new elementary proof of the Lucas–Lehmer test for Mersenne primes.

To echo the words of G.H. Hardy and E.M. Wright in the preface to their wonderful book, *An Introduction to the Theory of Numbers*, first published in 1938: the material of elementary number theory is so attractive that only extravagant incompetence could make it dull. I trust, therefore, that I have failed to make the material dull; in fact I dare to hope that combining this beautiful mathematics with fairly simple computing makes the subject even more attractive than before.

Peter Giblin

Logical dependence of chapters

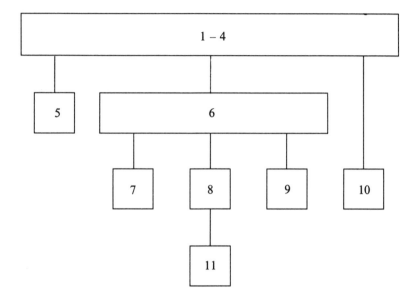

1

The fundamental theorem, greatest common divisors and least common multiples

1. Primes and the fundamental theorem

This book has for its theme the identification of prime numbers and the factorization of numbers known to be composite. We say one integer a is a *factor* of another integer b (written $a \mid b$ and also pronounced 'a divides b' or 'a is a divisor of b') if $b = ac$ for some integer c. Let us consider positive integers for a moment. A (positive) *prime* is an integer $p > 1$ which has no positive factor besides 1 and p, while a (positive) *composite* is an integer $n > 1$ which has 'nontrivial' factors: $n = ab$ where $a > 1$ and $b > 1$. (Thus note in passing that 1 is neither prime nor composite; this saves trouble later.)

The emphasis in this book is on number theory which can be illustrated by computer programming, but in the first few sections we shall concentrate on setting up the most basic facts about whole numbers (though some of our illustrations will be less basic!). Our starting point is the fact that all whole numbers > 1 can be factorized into products of primes, the product being unique up to reordering of the factors. For example, $1000 = 2^3 \cdot 5^3$, $119 = 7 \cdot 17$, $997 = 997$ (already prime). The general statement is as follows.

1.1 Fundamental theorem of arithmetic *Every integer $n > 1$ has a factorization into primes: $n = p_1^{n_1} p_2^{n_2} \ldots p_k^{n_k}$, $p_1 < p_2 < \ldots < p_k$ primes and $n_1, \ldots, n_k$ integers > 0, while $k \geq 1$. Furthermore the p_i and the n_i are uniquely determined by n.*

Many books do not prove this theorem until rather later, basing their proof on properties of primes, for example [Hardy and Wright (1979), or any previous edition], [K.H. Rosen (1988), p.85], [Niven *et al.* (1991), p.23]. There is a proof (due to F.A. Lindemann (1933)) which does not use later results; Lindemann's proof is given as an alternative in the books of Hardy and Wright and of Niven *et al.* cited above, and also in [Davenport (1982)]. For completeness we give this proof at the end of the section, as 1.8 below, but you may wish to take the fundamental theorem on trust!

The factorization in 1.1 is referred to as the *prime-power* factorization of n. According to 1.1, the number 1 does not have a prime-power factorization; we can give it an artificial one by taking $k = 0$ (the empty factorization!) and declaring that the right-hand side is then 1. This artificiality will not cause any problems. The number 0 definitely does not have a prime-power factorization!

Note the immediate consequences:

1.2   *A prime p is a factor of $n \Leftrightarrow p$ is one of the p_i occurring in 1.1. If p and q are distinct primes and $p \mid n$, $q \mid n$, then $pq \mid n$. The same holds for three or more distinct primes.*

Of course negative numbers can be factorized too: we factorize $n < -1$ by factorizing $-n$ and putting a minus sign in front, for example $-200 = -2^3 \cdot 5^2$. Written in this way the factorization into powers of positive primes, with a minus sign in front, is still unique. An integer $p < 0$ is declared to be prime if and only if $-p$ is prime, but from now on the unqualified word 'prime' will mean 'positive prime'. Perhaps the 'correct' way to think of all this is to say that 1 and -1 are the 'units' in $\mathbf{Z}$, the set of integers, and that p is a prime if and only if it is not a unit and every factorization of p, $p = ab$ with a and $b \in \mathbf{Z}$, has a or b a unit.

When we have two numbers a and b to consider it is often convenient to use the same primes in their decompositions, by relaxing the $n_i > 0$ condition to $n_i \geq 0$: note that uniqueness in 1.1 then applies only to the primes occurring with power > 0 and we include only primes occurring in a or b or both. For example,

$$28 = 2^2 \cdot 5^0 \cdot 7^1 \text{ and } 200 = 2^3 \cdot 5^2 \cdot 7^0.$$

If $a = p_1^{a_1} p_2^{a_2} \ldots p_k^{a_k}$ and $b = p_1^{b_1} p_2^{b_2} \ldots p_k^{b_k}$ we have $ab = p_1^{a_1+b_1} p_2^{a_2+b_2} \ldots p_k^{a_k+b_k}$ and $a_i + b_i > 0$ for each i. Let p be prime. Then $p \mid ab$ if and only if p is one of the p_i. Hence:

1.3 *If p is prime and p | ab then p | a or p | b (or both).*

This very important fact must *never* be misread as saying that, for any n, if $n \mid ab$ then $n \mid a$ or $n \mid b$. Just try $4 \mid 2 \cdot 6$. In 1.3, primality is all. While we're about it, let us note one other immediate consequence of 1.1:

1.4 *a | b if and only if every prime occurring in the prime-power decomposition of a occurs also in that of b, and the power in a is $\leq$ that in b.*

The fundamental theorem 1.1 has a simple look about it, and indeed it *is* nice to know that every number > 1 either is prime ($k = 1$, $n_1 = 1$) or factorizes into primes. But actually *finding* the factorization, or *proving* the primality can be very hard (otherwise this book would have been much shorter than it is). We pause here to give examples, at least one of which is rather painful. Note that a 'factorization' of a number n is just an expression $n = ab$ where $a > 1$ and $b > 1$. The *complete factorization* of n is the one in 1.1, and that may be much harder to establish.

The great number theorist P. de Fermat (1601–1665) conjectured that all numbers of the form $2^{2^n} + 1$ might be prime, having tested $n = 0, 1, 2, 3$ and 4. In 1732, L. Euler (1707–1783) disproved the conjecture by showing that $2^{32} + 1 = 641 \times 6\,700\,417$. (Considering 641 is only the 116th prime it is maybe very slightly surprising that Fermat didn't discover this for himself!) Since that time a veritable cottage industry has arisen for factorizing the 'Fermat numbers' $F_n = 2^{2^n} + 1$. $F_6 = 274\,177 \times 67\,280\,421\,310\,721$ was factorised by F. Landry in 1880. M.A. Morrison and J. Brillhart completely factorized F_7 as recently as 1970, F_8 had to wait until 1980 (R.P. Brent and J.M. Pollard) and F_9 (with 155 digits) until 1990 (M. Manasse and A. Lenstra; see [Cipra (1990)]). We shall study the method which was used to factorize F_7 in Chapter 10. For the method used to factorize F_8, see Chapter 3, 3.6.

Whether the sequence will continue with F_{10} (which has 309 digits) being factorized in the year 2000 must remain doubtful; it is estimated that it would need half a million times the resources which were used to factorize F_9, and that consisted of 1000 computers around the world working for two months! But, on the other hand, someone may come up with an even cleverer method for factorization.

In both the case of F_9 and that of F_{10} one 'small' factor has been known for some time; compare Chapter 4, 2.4(e) and (f). It is worth noting that there is little chance of completely factorizing such large numbers by the naive method of testing for divisibility by the primes $2, 3, 5, 7, 11, \ldots$ until a factor is found. If the prime factors are large then even on the most optimistic estimates of computational speed, it would take far longer than the age of the universe to factorize by trial division. We take up this question in slightly more detail in Chapter 2, 3.5 below. An extraordinary example of a factor being known of an inconceivably vast number is the result of W. Keller in 1984 that $5 \cdot 2^{23\,473} + 1$ is a factor of the Fermat number $F_{23\,471}$. Maybe you can verify that this Fermat number has more than 10^{7000} *digits!*

Besides factorizing F_5 Euler also proved (1772) that $2^{31} - 1 = 2\,147\,483\,647$ is prime; you might like to consider how you would do that for yourself. (Presumably you would try dividing the number by one prime number after another in increasing order of magnitude. When would you stop and declare that $2^{31} - 1$ is prime?) For a century or so afterwards, this was the largest prime known. All the largest primes known today are also of the form $2^n - 1$ and are known as *Mersenne primes* after M. Mersenne (1588–1648) who first studied them and corresponded about them with other mathematicians including Fermat. See the Index for other references to Mersenne primes.

Here is a more painful story. According to [Kraitchik (1952)] the mathematician F. Landry 'devoted his life to the investigation of numbers of the form $2^n \pm 1$'. The one which gave him most trouble was $n = 58$, which, apart from an easy factor of 5, has two very large factors which turn out to be prime:

$$2^{58} + 1 = 5 \times 107\,367\,629 \times 536\,903\,681.$$

Landry discovered this in 1869. In 1871, A. Aurifeuille pointed out that

$$2^{58} + 1 = (2^{29} - 2^{15} + 1)(2^{29} + 2^{15} + 1),$$

so that it was a mistake to take out that easy factor of 5! Much is known about numbers $2^n \pm 1$; see for example [Kraitchik (1952), Brillhart and Selfridge (1967), Riesel (1987)].

A happier tale is told of F.N. Cole, who, at a meeting of the American Mathematical Society in October 1903 gave a silent lecture in which he verified by hand that

$$2^{67} - 1 = 193\,707\,721 \times 761\,838\,257\,287.$$

The audience was enthusiastic and so far no-one has discovered a trivial reason for this factorization.

There is a wonderful account in [Sacks (1985), Chap. 23] of autistic twins John and Michael, who were able to recognize large primes (of 6 to 10 digits) by 'some unimaginable internal process of testing'. These 'prime twins' should not be confused with 'twin primes', of which more later.

1.5 Exercises

(a) Verify that $2^{4n+2} + 1 = (2^{2n+1} - 2^{n+1} + 1)(2^{2n+1} + 2^{n+1} + 1)$. By taking $n = 4$ show that both factors may be composite. How can this formula be regarded as a special case of the rather surprising factorization $x^4 + 4y^4 = (x^2 - 2xy + 2y^2)(x^2 + 2xy + 2y^2)$? Note that in particular this proves $x^4 + 4$ is composite for all $x > 1$.

(b) Verify that $2^{3n} \mp 1 = (2^n \mp 1)(2^{2n} \pm 2^n + 1)$. Does this sometimes give a different factorization from (a)?

(c) Verify that $2^{mn} - 1 = (2^m - 1)(2^{m(n-1)} + 2^{m(n-2)} + \ldots + 1)$ and $a^n - 1 = (a - 1)(a^{n-1} + a^{n-2} + \ldots + 1)$. Suppose now that $n \geq 2, a \geq 2$ and $a^n - 1$ is prime. Deduce that $a = 2$. Show further that n is prime (e.g. by considering the special case $a = 2^m$). Such primes $2^n - 1$, for n prime, are, as noted above, called 'Mersenne primes'.

(d) There are more spectacular algebraic factorizations due to E. Lucas (see [Riesel 1987, Appendix 6]), for example $a^{10} + 1 = (a^2 + 1)((a^4 + 5a^3 + 7a^2 + 5a + 1)^2 - 10a(a^3 + 2a^2 + 2a + 1)^2)$. The point of this is that, if $10a$ is a perfect square, then the second factor is a difference of two squares and so factorizes further. Verify the formula and try some examples, such as $a = 10^3$.

(e) Show that, for any $k \geq 2$, the $k-1$ numbers $k!+2, k!+3, \ldots, k!+k$ are all composite. Thus the gaps between successive primes can be arbitrarily large. For example when $k = 8$, the numbers $40\,322, \ldots, 40\,328$ are all composite. But these are not the smallest 7 consecutive composite numbers. Find these numbers (they are all less than 100). Check that the first 13 consecutive composite numbers start with 114. Gaps between consecutive primes will be studied in Chapter 2, 1.6, 1.9 below; needless to say the first occurrence of gaps of a given size has been investigated by computer. A recent reference is [Young and Potler (1989)]. We shall have occasion to mention the distribution of prime numbers on several occasions in this book, but the detailed mathematics is usually too technical to include. See 'prime number theorem' in the index for references.

(f) Let n be an odd composite positive integer. Show that $(2^n + 1)/3$ is odd and composite. (See also *Math. Gaz.* **78** [1994], 167–172.)

(g) Let n be composite. Show that $(10^n - 1)/9$ is composite. Note that this is $10^{n-1} + 10^{n-2} + \ldots + 10^1 + 1$, so *is* an integer! It has decimal representation consisting of n ones and is known as a *repunit*. In fact not many prime repunits are known apart from $n = 2$. The result here shows that for the repunit to be prime, n must be prime; try $n = 3$ and 5 to see that the converse is not true.

(h) Suppose that a and c are all integers > 1. Show that $(2^{ab} - 1) = (2^a - 1)(2^c - 1)$ is impossible. Deduce that, if n has two distinct proper divisors (i.e. distinct divisors $\neq \pm 1$ and $\pm n$) then $2^n - 1$ cannot be the product of two primes. [Show $2^n - 1$ has at least three distinct proper divisors; compare the first formula of (c) above.] Show similarly that if $n = c^k$ for $c \geq 2, k \geq 3$, then $2^n - 1$ cannot be the product of two primes. What does this show about values of n for which $2^n - 1$ is the product of two primes?

(i) Let p be a prime and k an odd integer ≥ 1, and suppose that $p + 1 = 2^t s$ where s is odd. Show that 2^t, and no higher power of 2, divides $p^k + 1$. [Hint. Use the binomial theorem.]

It has been known for well over 2000 years that there are infinitely many prime numbers. The proof given by Euclid in Book IX, Proposition 20 of the *Elements* is one of the most famous in all of mathematics.

1.6 Theorem   *There are infinitely many prime numbers.*

Proof   There is certainly at least one prime number, namely 2. Now let $p_1, p_2, \ldots, p_k$ be primes and consider $Q = p_1 p_2 \ldots p_k + 1$. By 1.1, Q has a prime factor but clearly it cannot be any of the p_i since they leave remainder 1 when divided into Q. So there must be another prime besides the k already listed. This applies for any k, so there are infinitely many primes.   □

Note that Euclid actually speaks of the 'least common measure' ($=$ lcm) of $p_1, p_2, \ldots, p_k$ but this equals their product as they are distinct primes.

As a means of discovering new primes, the procedure of the proof is less than helpful. For example $2 \cdot 3 \cdot 5 \cdot 7 \cdot 11 + 1 = 2311$ actually is prime, and adding that to the list $2 \cdot 3 \cdot 5 \cdot 7 \cdot 11 \cdot 2311 + 1 = 5\,338\,411$ which happens to be $13 \cdot 19 \cdot 21\,613$. So 13 arrives via 2311. No-one knows whether, taking the *first k* primes for the p_i, the number Q is prime for infinitely many k, or indeed composite for infinitely many k Apparently [Ribenboim (1988) p. 4] it is known that Q is prime for $k = 1, 2, 3, 4, 5, 11$ and a handful of other values, but most 'small' values of k give a composite Q.

Here is a variant of Euclid's proof which gives a stronger result.

1.7 Theorem   *There are infinitely many primes of the form $4r - 1$.*

Proof   Note that all primes are of the form $4r + 1$ or $4r - 1$, apart from 2, and certainly both kinds exist. Let $p_1, p_2, \ldots, p_k$ be primes of the form $4r - 1$, none of them equal to 3, and consider $Q = 4p_1 p_2 \ldots p_k + 3$. Now Q factorizes into primes, none of which can be 2 since Q is odd. If all the primes were of the form $4r+1$ then, multiplying them all together to get Q, the result would also be of the form $4r+1$, whereas Q is certainly not of this form. Hence at least one prime dividing Q has the form $4r-1$. As in Euclid's proof it cannot be any of the p_i (nor can it be 3) and, as before, this establishes the infinitude of primes of the form $4r-1$.   □

We shall meet more sophisticated arguments later which establish that there are infinitely many primes of other special forms such as $4r+1$. An argument similar to 1.7 does not work here.

*1.8 Proof of the fundamental theorem 1.1   First, we show that a factorization exists. Starting with $n>1$, if n is prime then we are finished; otherwise $n=ab$ where $a>1$ and $b>1$. Apply the same argument to a and to b: each either is prime or, if not, is a product of two numbers both >1. The numbers other than primes involved in the expression for n are >1 and decrease at every step; hence eventually all the numbers must be prime.

Now comes the hard part: uniqueness. Suppose that the theorem is false and let the *smallest* number >1 for which the theorem fails be n. Thus

$$n = p_1p_2\ldots p_r = q_1q_2\ldots q_s,$$

where the ps and qs are prime. Clearly both r and s must be >1 (otherwise n is prime, or a prime is equal to a composite). If for example p_1 were one of the qs, then n/p_1 would have two expressions as a product of primes, but $n/p_1 < n$ so this would contradict the definition of n. Hence p_1 is not equal to any of the qs and similarly none of the ps equals any of the qs. We may suppose $p_1 < q_1$; define

$$N = (q_1 - p_1)q_2q_3\ldots q_s = p_1(p_2p_3\ldots p_r - q_2q_3\ldots q_s).$$

Certainly $1 < N < n$, so N is uniquely factorizable into primes. However $p_1 \nmid (q_1 - p_1)$, since $p_1 < q_1$ and q_1 is prime; hence the above expressions for n have one containing p_1 and the other not. This contradiction proves the result: there cannot be any exceptions to the theorem.   □

2. Greatest common divisor and least common multiple

Suppose we are given two positive integers a and b; how can we determine their *common divisors*, that is, the numbers d for which $d \mid a$ and $d \mid b$? In principle the answer is given by 1.1: if $a = p_1^{a_1}p_2^{a_2}\ldots p_k^{a_k}$, $b = p_1^{b_1}p_2^{b_2}\ldots p_k^{b_k}$, with the a_i and $b_i \geq 0$, then $d \mid a$ and $d \mid b$ if and only if $d = p_1^{d_1}p_2^{d_2}\ldots p_k^{d_k}$ where $d_i \leq a_i$ and $d_i \leq b_i$, that is, $d_i \leq \min(a_i, b_i)$. Clearly the *greatest* of these common divisors is the common divisor h

given by $d_i = \min(a_i, b_i)$ for all i. If a or b is a negative integer then we can insert a minus sign before one or other prime-power decomposition and a $\pm$ sign before that of d; the greatest value of d is then with a $+$ sign and the same d_i as before. Note that if say $a = 1$ then all the a_i are 0 and $h = 1$.

2.1 Definition   *The common divisor h just mentioned, where the power of p_i is $\min(a_i, b_i)$, is called the* greatest common divisor (gcd) *of a and b. This is also known as the* highest common factor (hcf) *of a and b. Note that it has the property that every common divisor is a factor of h: if $d \mid a$ and $d \mid b$ then $d \mid h$. The notation is $h = \gcd(a, b)$ or just $h = (a, b)$. Two integers a and b with $(a, b) = 1$ are called* coprime *or* mutually prime *or* relatively prime.

Of course $(a, b) = (b, a)$ and $(a, 1) = |a|$ for any nonzero a and b. If we want a meaning for $(a, 0)$ it is usually taken to be the absolute value $|a|$ of a, unless $a = 0$ when there is no meaning assigned. Here are some useful properties of the gcd.

2.2 Proposition
   (a) *If p is prime then $(p, a) = p$ when $p \mid a$ and $(p, a) = 1$ otherwise. Also if $p \nmid a$ then $(p^n, a) = 1$ for any $n \geq 1$.*
   (b) *If $(a, b) = h$ then $(a/h, b/h) = 1$.*
   (c) *If $a \mid bc$ and $(a, b) = 1$ then $a \mid c$. [Note. The hypothesis $(a, b) = 1$ cannot be weakened to $a \nmid b$. Try $a = 4, b = 2, c = 6$.]*
   (d) *If p is prime, $p^n \mid ab$ and $p \nmid (a, b)$, then $p^n \mid a$ or $p^n \mid b$. [Note. This is false without $p \nmid (a, b)$, e.g. $2^2 \mid 6 \cdot 10$ but $2^2 \nmid 6$ and $2^2 \nmid 10$.] If $(a, b) = 1$ then the hypothesis $p \nmid (a, b)$ can be dropped.*

Proof   These are all very easy to prove from the prime-power decompositions of the numbers involved. Let us prove (c) by this method and leave (a) and (b) as exercises. Thus we have $a = p_1^{a_1} p_2^{a_2} \ldots p_k^{a_k}$, $b = p_1^{b_1} p_2^{b_2} \ldots p_k^{b_k}$, $c = p_1^{c_1} p_2^{c_2} \ldots p_k^{c_k}$, where a_i, b_i, c_i are all ≥ 0. (We leave to the reader insertion of $-$ signs if a or c is < 0.) Since $(a, b) = 1$ we have, for each i, a_i or $b_i = 0$. Since $a \mid bc$ we have, for each i, $a_i \leq b_i + c_i$, and then either possibility gives $a_i \leq c_i$, which proves that $a \mid c$.

Let us deduce (d) from the other parts. Thus, if $p^n \mid ab$ then $p \mid ab$, so that, since p is prime, $p \mid a$ or $p \mid b$, but not both since $p \nmid (a, b)$. If $p \nmid a$, then $(p^n, a) = 1$ by (a), so, by (c), $p^n \mid b$. $\square$

2.3 Exercises
   (a) Show that if $(a, b) = 1$ and $(a, c) = 1$, then $(a, bc) = 1$.
   Show that, if $a \mid n$, $b \mid n$ and $(a, b) = 1$, then $ab \mid n$. [This can be done

directly from prime-power factorizations, or by writing $ra = sb = n$ say
and using $(a, b) = 1$ to show $a \mid s$.]

Show that, if $a \mid b$ and $(b, c) = 1$, then $(a, c) = 1$.

(b) Show that if $a \mid b$ then $(a, b) = |a|$.

(c) Show that if $(a, b) = 2$ and $a \mid k$, $b \mid k$ then $ab \mid 2k$. Give an
example to show that ab may not divide k.

(d) Show that, if $1 \le r \le p - 1$ where p is prime, then $(r!, p) = 1$.
Deduce that the binomial coefficient

$$\binom{p}{r} = \frac{p!}{r!(p - r)!} = \frac{p(p - 1) \dots (p - r + 1)}{r!}$$

is a multiple of p.

Supposing that n is not a prime, say $n = kp^s$ where p is prime, $p \nmid k$
and also $k > 1$ or $s > 1$, then $\binom{n}{p}$ is *not* divisible by n, i.e., *some*
nontrivial binomial coefficient is not a multiple of n.

(e) Suppose a, b, x, y are nonzero integers and $x/a = y/b$ (these frac-
tions will not necessarily be integers!). Let $h = (a, b)$ and let $a = a_1 h, b = b_1 h$ for integers a_1, b_1. Use 2.2(c) to show that $x = ka_1, y = kb_1$ for some
integer k.

(f) Let a, b, c, d be > 0 and $(a, b) = 1, (c, d) = 1$, so that the fractions
$\frac{a}{b}$ and $\frac{c}{d}$ are in their lowest terms. Suppose that $\frac{a}{b} + \frac{c}{d}$ is an integer.
Show that $b = d$.

(g) As an application of (e), consider an $a \times b$ rectangle (a, b positive
integers) divided into ab squares by a grid of lines parallel to the edges

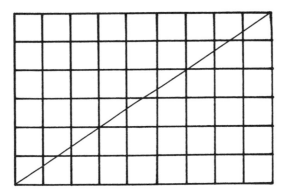

Fig. 1.1.

and draw a diagonal of the rectangle (Fig. 1.1). Show that the diagonal
passes through precisely the intersections of grid lines with coordinates
$x = ka_1, y = kb_1$ for $k = 0, 1, 2, \dots, h$. Deduce that the diagonal passes
across precisely $a + b - (a, b)$ grid squares.

(h) Given three nonzero integers a, b and c, show from prime-power factorizations that $((a, b), c) = ((b, c), a) = ((c, a), b)$. The common value of these three is written (a, b, c) and called the gcd of a, b and c. Show that $((a, b), (a, c)) = (a, b, c)$.

(i)* Let a, b, c, x, y, z be nonzero integers and suppose that $x/a = y/b = z/c$. Let $h = (a, b, c)$ and $a = a_1 h$, $b = b_1 h$, $c = c_1 h$. Show that $(a_1, b_1, c_1) = 1$. Using such facts as $((a_1, b_1), (a_1, c_1)) = 1$, and (a) above, deduce that $x = k a_1$, $y = k b_1$, $z = k c_1$ for an integer k.

(j)* Consider an $a \times b \times c$ rectangular brick, divided into abc unit cubes by planes parallel to the faces, and a diagonal of the brick (passing through the interior). Use (i) to show that this diagonal passes through precisely $(a, b, c) + 1$ corners of the unit cubes. Using (e), show that the number of unit cubes which the diagonal traverses is $a + b + c - ((a, b) + (b, c) + (c, a)) + (a, b, c)$. [This formula generalizes to n-dimensional bricks, and is an example of the 'principle of inclusion–exclusion' which we shall meet in connexion with Legendre's sieve in Chapter 2, 4.4 below.]

(k) Let p_i denote the ith prime ($p_1 = 2$, $p_2 = 3$, etc.) and let $p_k < n < p_{k+1}^2$ (for some $k \geq 1$). Let m denote the product of the first k primes and let $m = qn + r$ where $0 \leq r < n$ (so r is the remainder on division of m by n). (For example, $7 < 101 < 11^2 (k = 4)$ and $m = 210 = 2 \cdot 101 + 8$.) Assume that $r \neq 0$. Show that n is prime if and only if $(n, r) = 1$. [The proof that if $(n, r) = 1$ then n is prime involves one subtlety which we shall exploit to the full later. It is enough to show that n is not divisible by any of $p_1, \ldots, p_k$, for, if n is composite, then *some* prime $\leq \sqrt{n}$ must be a factor of n. See 1.1 of Chapter 2, below, if you don't see why this is.] For more discussion of this result, see [Guy, et al. (1987)].

(l) Verify that the numbers 2184 to 2200 inclusive have the property that any one of them has a common factor > 1 in common with at least one of the others. Show that this property fails if the list is extended at either end. [It is an unusually long list for the size of numbers involved!]

There is a simple property of factors of products of two numbers which is worth inserting here; we shall use it in Chapter 9. All numbers occurring here are positive.

2.4 Proposition   *Suppose that $(a, b) = 1$ and that $d \mid ab$. Then there exist unique numbers d', d'' with $d = d'd''$, $d' \mid a$, $d'' \mid b$. It is automatically true that $(d', d'') = 1$. Conversely, if $d' \mid a$ and $d'' \mid b$, then $d'd'' \mid ab$, so divisors of ab are in one-to-one correspondence with pairs of divisors of a and b.*

Proof   The idea is that d' is the 'part of d which goes into a' and d''

is the 'part of d which goes into b'. With the usual decompositions, $a = p_1^{a_1} p_2^{a_2} \ldots p_k^{a_k}$, $b = p_1^{b_1} p_2^{b_2} \ldots p_k^{b_k}$, we have a_i or $b_i = 0$ for each i, and $d = p_1^{d_1} p_2^{d_2} \ldots p_k^{d_k}$ where

$$d_i \le a_i + b_i = \begin{cases} a_i & \text{if } b_i = 0, \\ b_i & \text{if } a_i = 0. \end{cases} \tag{1}$$

Now d' must be a product of powers of those p_i where $b_i = 0$ (since $d' \mid a$), and d'' must be a product of powers of those p_i where $a_i = 0$ (since $d'' \mid b$). In order to obtain $d = d'd''$ we must then take $d' = \prod p_i^{d_i}$, with the product over all i with $b_i = 0$ (so $d' \mid a$ by (1)), and $d'' = \prod p_i^{d_i}$ with the product over all i with $a_i = 0$ (so $d'' \mid b$ by (1)). Since d' and d'' involve distinct sets of primes, $(d', d'') = 1$.

Let, as before, $a = p_1^{a_1} p_2^{a_2} \ldots p_k^{a_k}$, $b = p_1^{b_1} p_2^{b_2} \ldots p_k^{b_k}$, with the a_i and $b_i \ge 0$. If we take the product of prime powers $\ell = p_1^{\ell_1} p_2^{\ell_2} \ldots p_k^{\ell_k}$ with $\ell_i = \max(a_i, b_i)$ we obtain a number ℓ which is a *common multiple* of a and b. Furthermore any multiple of both a and b must have these primes to at least these powers, so is a multiple of ℓ: if $a \mid m$ and $b \mid m$ then $\ell \mid m$. The number ℓ is called the *least common multiple* of a and b, denoted $\mathrm{lcm}(a, b)$ or $[a, b]$. (If a or b is < 0 we still take $[a, b] > 0$: it is the least (positive) common multiple.) Since $\max(a_i, b_i) + \min(a_i, b_i) = a_i + b_i$ the following result is immediate.

2.5 Proposition   *For nonzero a, b* $\mathrm{lcm}(a, b) \times \gcd(a, b) = |ab|$.   □

2.6 Exercises

(a) Of course, if $h = (a, b)$ and $\ell = [a, b]$, then $h \mid \ell$. Conversely if $h \mid \ell$, then there exist a, b (namely $a = h$ and $b = \ell$) for which $h = (a, b)$ and $\ell = [a, b]$. Find *all* a and b for which $(a, b) = 12$ and $[a, b] = 72$. What is the general procedure here? When will the solution be unique?

(b) Let r and s be integers. Show that they can be written $r = ux$, $s = vy$ where $(u, v) = (x, y) = 1$ and $uv = \mathrm{lcm}(r, s)$, $xy = \gcd(r, s)$, by defining x and y as follows. Let $r = p_1^{r_1} p_2^{r_2} \ldots p_k^{r_k}$, $s = p_1^{s_1} p_2^{s_2} \ldots p_k^{s_k}$ as usual; then define $x = p_1^{x_1} p_2^{x_2} \ldots p_k^{x_k}$, $y = p_1^{y_1} p_2^{y_2} \ldots p_k^{y_k}$ where

$$x_i = \begin{cases} r_i & \text{if } r_i < s_i, \\ 0 & \text{otherwise,} \end{cases} \quad \text{and} \quad y_i = \begin{cases} s_i & \text{if } r_i \ge s_i, \\ 0 & \text{otherwise.} \end{cases}$$

Note that when $r_i = s_i$ we could put the corresponding power of p_i into either x or y while still preserving the required properties. Find all possible solutions for u, v, x and y, including the suggested one, when $r = 100$ and $s = 60$; and when $r = 700$ and $s = 300$.

(c) Show by prime-power factorizations that $[[a, b], c] = [[b, c], a] = [[c, a], b]$. The common value of these is written $[a, b, c]$ and called the lcm of a, b and c. Of course this generalizes to $[a_1, a_2, \ldots, a_n]$.

(d) Let $q_1 < q_2 < \ldots < q_8$ be eight odd primes and let r be the lcm of all the numbers $q_j - q_i$ for $1 \leq i < j \leq 8$. Let $A = \{5, 7, 8, 9\}$. Show that dividing any odd prime by an element of A can give a maximum of seven different remainders. (For instance, $q/7$ can have remainders $0, 1, \ldots, 6$ and $q/8$ can have remainders $1, 3, 5, 7$ since q is odd.) It follows that, for each $m \in A$, at least two of the q_i must leave the same remainder when divided by m, and consequently their difference is divisible by m. (This is an application of the famous 'pigeonhole principle', the pigeons here being the q_i and the pigeonholes the remainders.) Why does it follow that r is divisible by 5, 7, 8 and 9 and hence by $5 \cdot 7 \cdot 8 \cdot 9 = 2520$? (Compare 2.3(a)). Check that the eight primes $11, 17, 23, 29, 41, 47, 53, 59$ give $r = 5040$. (In fact r is always divisible by 2520. See [Morain (1989), (1990)].)

(e) Show that $\max(i, j, k) = i + j + k - \min(j, k) - \min(k, i) - \min(i, j) + \min(i, j, k)$ for any integers i, j and k. [Note that by symmetry it is enough to check the formula when $i \leq j \leq k$.] Deduce that for nonzero integers a, b and c,

$$[a, b, c] = \frac{abc(a, b, c)}{(b, c)(c, a)(a, b)}.$$

3. Euclid's algorithm for the gcd

The snag with the above approach to greatest common divisors is that the prime-power decompositions of a and b may be very hard to determine. Fortunately the situation is saved by the following extremely simple fact.

3.1 Proposition   *For any integer k, $(a, b) = (a + kb, b)$.*

Proof   The simplest way to prove this is to show that the set of common divisors of $a + kb$ and b coincides with the set of common divisors of a and b, that is, $d \mid a$ and $d \mid b \Leftrightarrow d \mid (a + kb)$ and $d \mid b$. This fact is more or less immediate from the definition of divisor. It certainly has the consequence that the *greatest* common divisor of a and b equals that of $a + kb$ and b.   □

3.2 Exercises

(a) Use 3.1 to show that $(6n + 1, 6n - 3) = 1$ for any integer n. [Take $k = -1$.] Also show $(5n + 3, 3n + 2) = 1$, perhaps by starting with $(5n + 3, 3n + 2) = (2n + 1, 3n + 2)$.

(b) Let p be prime and $p \mid n$, where p and n are both odd and $p < n$. Let $a = (n + 1, p + 1)$, $b = (n - 1, p - 1)$. Show that $(a, b) = 2$, $(a, p) = 1$,

$(b, p) = 1$. Deduce from 2.3 (a) and (c) (beware different notation!) that $abp \mid 2(n - p)$ and hence that $ab < 2n/p$.

(c) *The division property*, or *division algorithm*. Let a and b be integers with $b \neq 0$. Then there exist *unique* integers q and r (the *quotient* and the *remainder*) with

$$a = bq + r, \quad 0 \leq r < |b|.$$

Prove this by the following method. Consider the set of integers

$$S = \{a - bq : q \in \mathbf{Z}\}.$$

Clearly S contains nonnegative integers (why?). Let r be the *smallest* integer in S which is ≥ 0. Show that r satisfies $0 \leq r < |b|$ (consider $b > 0$ if you prefer). To prove q and r are unique, suppose $a = bq + r = bq_1 + r_1$ where $0 \leq r < |b|$ and $0 \leq r_1 < |b|$. Assuming $r < r_1$ deduce $0 < r_1 - r < |b|$ and use this and $r_1 - r = b(q - q_1)$ to get a contradiction. Deduce $r_1 = r$ and hence $q_1 = q$.

Possibly any two numbers can be reduced, in the manner of 3.1, to two simple numbers whose gcd is immediately recognizable? This is indeed so, and is the content of Euclid's algorithm. (Incidentally the word *algorithm* comes from the name of the Arabic mathematician al-Khowârizmî. He lived some 1100 years after Euclid put the gcd algorithm as Proposition 2 in Book VII of his *Elements*, written around 300 B.C. Euclid's algorithm is intimately connected with the subject of continued fractions, which we shall take up in Chapter 10. Much has been made of the possible implications of this connexion for the interpretation of ancient Greek mathematics; see [Fowler (1987)]. The first use of the term 'algorithm' with its modern meaning appears to have been in the first edition of G.H. Hardy and E.M. Wright's famous book *An Introduction to the Theory of Numbers* (1938).)

We use the *division property* of integers proved, but for a few details, in 3.2(c) above. We usually apply the division when $b > 0$ and then the non-negative remainder r is *less than the number you divide by*.

3.3 Euclid's algorithm for the gcd of a and b
Let a and b be integers with $b > 0$ (since $(a, b) = (-a, -b)$ this is no loss of generality). We apply the division property repeatedly, as follows

(an example is given to the right of the general statement).

$$a = bq_1 + r_2, \qquad 0 \le r_2 < b; \qquad 330 = 140 \cdot 2 + 50;$$
$$b = r_2 q_2 + r_3, \qquad 0 \le r_3 < r_2; \qquad 140 = 50 \cdot 2 + 40;$$
$$r_2 = r_3 q_3 + r_4, \qquad 0 \le r_4 < r_3; \qquad 50 = 40 \cdot 1 + 10;$$
$$\cdots \qquad\qquad \cdots \qquad\qquad 40 = 10 \cdot 4.$$
$$r_{n-2} = r_{n-1} q_{n-1} + r_n, \quad 0 \le r_n < r_{n-1};$$
$$r_{n-1} = r_n q_n.$$

The remainders $r_2, r_3, \ldots$ are all ≥ 0 and decrease steadily, so they must eventually become zero. Repeated application of 3.1 shows that $(a, b) = (r_2, b) = (r_2, r_3) = \ldots = (r_{n-1}, r_n) = r_n$, since in fact $r_n \mid r_{n-1}$. Hence *the gcd of a and b is the last nonzero remainder in the repeated division process above.*

In the example, $(330, 140) = 10$. Note that this was obtained without factorizing either number. What we *did* need to do was to divide one number into another to find the quotient and remainder. Fortunately computers are well set up to do this operation (we need not trouble ourselves with asking how they do it!). When $b > 0$, we have

$$a = bq + r, \quad 0 \le r < b \quad \text{implies} \quad \frac{a}{b} = q + \frac{r}{b}, \quad 0 \le \frac{r}{b} < 1.$$

Consequently q is just the 'whole number part' of a/b (it is what we think of as the 'number of times b goes into a'. More precisely, q is the next integer down from a/b, or a/b itself if this is an integer. This quantity is called the *integer part* of a/b:

3.4 Definition   *The* integer part *of a real number· x, denoted* $[x]$, *is the greatest integer* $\le x$.

Note   There is another common notation for the integer part, or *floor function*, namely $\lfloor x \rfloor$. This is to be contrasted with the *ceiling function*, denoted $\lceil x \rceil$, which is the smallest integer $\ge x$: the next integer *above* x, unless x is itself an integer in which case $\lceil x \rceil = x$. No use is made in this book of the ceiling function, and the integer part will be denoted by $[x]$ as in 3.4.

Thus $[1.6] = 1$, $[-1.6] = -2$, $[3] = 3$, $[-3] = -3$, etc. In the division $a = bq + r$, with $b > 0$, we have $q = [a/b]$. (The remainder r is sometimes called the 'least positive residue of a modulo b' and denoted $\langle a \rangle_b$, but we shall not use this terminology or notation yet.) We shall investigate the function $[x]$ in more detail below.

In BASIC, the function INT is exactly this, but in PASCAL the function INT 'rounds towards zero' so that $\text{INT}(-3.2) = -3$ for example. So long as we stick to positive numbers this is no problem. The repeated division of Euclid's algorithm for the gcd is realised by working out $r = a - b[a/b]$, that is $r = a - bq$, and then replacing a by b, b by r and repeating. Here is a PASCAL program for the algorithm.

3.5 Program for Euclid's algorithm with $a > 0, b > 0$

```
PROGRAM P1_3_5;
VAR
  a,b,r:  extended;
  (* Use type real if you do not have extended *)
  (* precision, but this limits the number of *)
  (* significant figures to about 11.  *)
BEGIN
  writeln ('type two positive numbers separated by a
  space');
  read (a,b);
  writeln ('gcd of', a:0:0, 'and', b:0:0, 'is');
  (* Since a and b change their values during the *)
  (* program, we have to put this here !*)
  WHILE b > 0 DO
    BEGIN
      r := a - b * INT (a/b);    (* See note 1 below *)
      a := b;
      b := r;
    END;
  writeln (a:0:0);    (* See Note 2 below *)
END.
```

Notes (1) If we allow $a < 0$ then the line indicated above must be replaced by the following (or something equivalent): `q := INT (a/b);`
`IF (q > a/b) THEN q := q - 1;`
`r := a - b * q;`
This is because INT in PASCAL rounds towards zero.
(2) When using real or extended type the 'formatting statement' `a:0:0` ensures that the answer is written without a decimal place, that is, as an integer.

3.6 Computing Exercise

Use the program to find gcd's of some small number pairs which you can also check by hand (this is always a good ploy when trying a new program!). Try also pairs such as $(666\ldots6, 777\ldots7)$ where there are k 6s and k 7s, so that the answer should be $111\ldots1$ with k 1s. At what value of k does the program start to give wrong answers?

3.7 Computing exercise: random pairs

You can produce random pairs of numbers by deleting the read statement and using

```
RANDOMIZE;
a := RANDOM(1000) + 1;
b := RANDOM(1000) + 1;
```

This produces random numbers between 1 and 1000. Run the program say 100 times and do a count to estimate the proportion of pairs which give (a) gcd $= 1$, (b) gcd $= 1$ or 2, (c) gcd > 10. (Putting a loop in the program to run it 100 times may not be satisfactory since RANDOMIZE tends to depend on the timer, and the program will run too fast for the time to change. A delaying loop can be used or a parameter such as RANDSEED(i*i).)

3.8 Computing exercise: modification of Euclid's algorithm

Consider the two lines $a = bq_1 + r_2$, i.e. $r_2 = a - bq_1$, and $b = r_2q_2 + r_3$, i.e. $r_3 = b - r_2q_2$. If a is replaced by $a - bq_1 = a - b[a/b]$, and *then* b is replaced by $b - aq_2 = b - a[b/a]$ (so long as the new $a \neq 0$!) then the final a and b are r_2 and r_3 respectively. Doing it again produces r_4 and r_5, and so on. Write a program which implements this modification. By incorporating a counting loop into 3.7 and the modification you can compare speeds of calculating gcd's of random pairs. Which program is faster?

3.9 Project: coprime pairs

Here is a 'heuristic argument' which suggests the proportion of pairs of numbers which should be coprime. We take all numbers to be $>$ 0. The 'probability' that a positive integer is divisible by a number h is presumably $1/h$, in the sense that, of the numbers $1, \ldots, n$, exactly $h, 2h, 3h, \ldots, [n/h]h$ are multiples of h, and $[n/h]/n$ will be close to $1/h$. Let p_h be the probability that two positive integers chosen at random have gcd $= h$. Note that $(a, b) = h \Leftrightarrow a/h, b/h$ are integers, and $(a/h, b/h) = 1$ (compare 2.2(b) above). So the probability that the gcd of two randomly chosen integers is h should be $p_h = (1/h)(1/h)p_1$. Since every pair of positive integers must have *some* gcd ≥ 1 we will then have

$$\sum_{h=1}^{\infty} p_h = 1, \text{ that is, } p_1 \sum_{h=1}^{\infty} \frac{1}{h^2} = 1.$$

The infinite sum $\Sigma(1/h^2)$ is known to be $\pi^2/6$ (see books on complex analysis, [Priestley (1990), p.131] for example), so that the probability p_1 of two randomly chosen numbers to be coprime is then $6/\pi^2 = 0.6079$ approximately. Using this result, what is the probability that two randomly chosen positive integers have gcd $= 1$ or 2? How about gcd > 10? Read an alternative heuristic argument for the value of p_1 in [Hardy and Wright (1979) §18.5 or Schroeder (1986), p.48] and write out your own account of it. Write a program which takes 1000 random pairs and counts the number which are coprime, which have gcd ≤ 2, and which have gcd > 10. Compare the results with the theoretical predictions. Also take *all* 2500 pairs a and b where $1 \leq a \leq 50$ and $1 \leq b \leq 50$ instead of a random collection of pairs.

3.10 Computing exercise
   Modify the program 3.5 (or 3.8) to calculate the gcd of three numbers a, b, c (compare 2.3(f) above).

   Returning to the example in 3.3 of a calculation of a gcd, namely $(330, 140)$, we can do the following calculation by starting at the end of the calculation and working upwards:

$$10 = 50 - 40 = 50 - (140 - 50 \cdot 2) = 50 \cdot 3 - 140 = (330 - 140 \cdot 2) \cdot 3$$
$$- 140 = 330 \cdot 3 - 140 \cdot 7.$$

The end result is that the gcd, 10, is expressed in the form $as + bt$, where in fact $s = 3$ and $t = -7$. Applying the same technique to the general algorithm in 3.3 we obtain the following result.

3.11 Proposition   *Let $(a, b) = h$. Then there exist integers s and t such that $as + bt = h$.*   □

   In fact the above method of calculating s and t is not ideal from a computational point of view since it starts at the end of Euclid's algorithm and works back to the beginning. We give a better method in 3.16 below, which also re-proves the result of 3.11. But first here are some theoretical consequences of 3.11.

3.12 Corollary   *$(a, b) = 1$ if and only if there exist integers s and t such that $as + bt = 1$.*

Proof $(a, b) = 1$ implies such s and t exist by 3.11. For the converse, if $as + bt = 1$ and $d \mid a$, $d \mid b$, then $d \mid (sa + bt)$ so $d \mid 1$, i.e. $d = \pm 1$. Hence the gcd of a and b is 1.   □

3.13 Exercises
   (a) Re-prove the result that if $a \mid bc$ and $(a, b) = 1$ then $a \mid c$ (2.2(c) above) by writing $as + bt = 1$ and mutliplying through by c.
   (b) Suppose that $(a, b) = 1$ and $c \mid (a + b)$. Show that $(c, a) = (c, b) = 1$.
   (c) Let $a = bq + r$, $0 \leq r < b$, as in the general step of the Euclidean

algorithm. Verify that, for $n > 1$ and $a > 0$,

$$n^a - 1 = (n^b - 1)(n^{b(q-1)+r} + n^{b(q-2)+r} + \ldots + n^{b+r} + n^r) + n^r - 1.$$

Write this as $A = BQ + R$. Why is $0 \leq R < B$? Deduce that the steps of the Euclidean algorithm for $(A, B) = (n^a - 1, n^b - 1)$ exactly follow those of the Euclidean algorithm for (a, b), replacing each r_i by $n^{r_i} - 1$. Deduce that, in particular, $(n^a - 1, n^b - 1) = n^{(a,b)} - 1$.

(d) Divide $(a^n - b^n)/(a - b) = a^n + a^{n-1}b + a^{n-2}b^2 + \ldots + b^n$ by $a - b$ and deduce that

$$\left(\frac{a^n - b^n}{a - b}, a - b \right) = (a - b, nb^{n-1})$$

Harder: use this result to show that the same gcd equals $(a-b, n(a,b)^{n-1})$. One possibility is to show that nb^{n-1} in the above equation can be replaced by $na^k b^{n-k}$ for $k = 1, 2, \ldots, n$. Then write $h = (a, b)$, $sa + tb = h$ and show that every common divisor of $a-b$ and nb^{n-1} also divides nh^{n-1}.

(e) Let $s_0 > a > 0$ be integers, $(s_0, a) = 1$ and define s_n for $n = 1, 2, 3, \ldots$ by $s_n = a + s_{n-1}(s_{n-1} - a)$. Show that

$$s_n - a = s_0 s_1 s_2 \ldots s_{n-1}(s_0 - a) \text{ for } n \geq 1.$$

Hence show that, if $n > m \geq 0$, then s_n and s_m are coprime.

As a particular case, take $s_0 = 3$, $a = 2$. Show by induction that $s_n = 2^{2^n} + 1$. (This is the nth Fermat number, which was mentioned in Section 1 above.)

(f) Consider the first step of Euclid's algorithm in the form $a = bq + r$, where we assume that $a > b > 0$, which implies $q \geq 1$. If $b \leq \frac{1}{2}a$ then of course $r < \frac{1}{2}a$. Assuming $b > \frac{1}{2}a$, show that r is still $< \frac{1}{2}a$. Why does this argument imply that, for any $i \geq 0$, $r_{i+2} < \frac{1}{2}r_i$ provided $r_i > 0$? (Here $r_0 = a$, $r_1 = b$.) Thus, so long as the remainders remain nonzero, we have $r_3 < \frac{1}{2}r_1$, $r_5 < \frac{1}{2}r_3$, $\ldots$. Deduce that $r_{2n+1} = 0$ as soon as n is large enough to make $2^n \geq b$. Try some examples to compare this upper bound on the length of the algorithm with the actual length.

(g) Let a and b be integers, not both zero. Consider the set of numbers $S = \{ax + by: x \text{ and } y \text{ integers}\}$. Let h be the smallest element of S which is > 0. By writing $a = qh + r$, $0 \leq r < h$, show that $r \in S$ and deduce that $h \mid a$ and similarly $h \mid b$. Show also that any common factor of a and b is a factor of h, and deduce that $h = (a, b)$. Note that this proves again the result of 3.11, but it is not a very useful way of actually finding integers s and t with $as + bt = (a, b)$.

(h) Use the result of (c) to show that, if $(a, b) = 1$, then $(r_a, r_b) = 1$ where r_a is the 'repunit' $(10^a - 1)/9$ which has decimal representation consisting of a ones. More generally, show that $(r_a, r_b) = r_{(a,b)}$.

(i) Let m and n be coprime integers and let u, v be integers > 0 with $nv - mu = 1$. (Why is it no loss of generality to take u and $v > 0$?) Verify that

$$x = a(a^m + b^m)^u, \quad y = b(a^m + b^m)^u, \quad z = (a^m + b^m)^v$$

is a solution to $x^m + y^m = z^n$. [The nonexistence of solutions when $m = n \geq 3$ is a famous conjecture usually called *Fermat's last theorem*.]

(j) Let p and q be distinct primes and let $n > 1$. Why do $n^p - 1$ and $n^q - 1$ both divide $n^{pq} - 1$? Use (c) above, and 2.4, to show that

$$N = \frac{(n^{pq} - 1)(n - 1)}{(n^p - 1)(n^q - 1)} \quad \text{is an integer.}$$

3.14 Exercise: linear diophantine equations

Consider the equation

$$ax + by = d \tag{1}$$

where a, b, d, x and y are integers and a, b are not both zero. We seek the complete set of solutions x, y of this equation. Let $(a, b) = h$, $a = a_1 h$, $b = b_1 h$ so that $(a_1, b_1) = 1$ (compare 2.2(b)).

(a) Show that, if $h \nmid d$, then (1) has no solution [$h \mid$ left-hand side of (1)...].

(b) Suppose now that $h \mid d$, $d = d_1 h$ say; then consider the equation

$$a_1 x + b_1 y = d_1 \tag{2}$$

Let s, t be chosen so that $a_1 s + b_1 t = 1$, so that $x = sd_1$, $y = td_1$ is one solution. Show that, if X, Y is any other solution, then $a_1(X - x) = -b_1(Y - y)$ and deduce that $X = x + kb$, $Y = y - ka$, for some integer k.

(c) Find the general solutions of each of the following equations.

$$7x + 5y = 17; \quad 6x + 9y = 33; \quad 6x + 12y = 36.$$

(d) Let a and b be > 0 and $a \nmid b$, $b \nmid a$. Show that integers x and y can be chosen so that $ax - by = (a, b)$ where $1 \leq x \leq b - 1$ and $1 \leq y \leq a - 1$.

(e) As an application of (d), consider a piece of paper 1 metre wide which is divided into a equal strips by $a - 1$ red lines, equally spaced and parallel ($a \geq 2$), and is also divided into b (≥ 2) equal strips by $b - 1$ blue lines, equally spaced and parallel to the red lines. Assuming that $a \nmid b$ and $b \nmid a$, show that the minimum nonzero distance between a red line and a blue line is $(a, b)/ab$ metres. [Recall from 3.13(g) that the gcd is the smallest positive number of the form $ax - by$.]

(f) Suppose we now consider the equation in integers

$$ax + by + cz = d \tag{3}$$

where we suppose that $(a, b, c) \mid d$, in order that the equation has any solutions. Recalling that $(a, b, c) = (a, (b, c))$ (see 2.3(h) above), we know from (b) that there exist x_0 and u_0 satisfying $ax_0 + (b, c)u_0 = d$, and also from (b) we know that there exist y_0 and z_0 satisfying $by_0 + cz_0 = (b, c)$. Find the general solution of $by + cz = k(b, c)$ for any fixed k, and hence obtain the general solution of (3) for arbitrary integers s and t:

$$x = x_0 + \frac{(b, c)}{(a, b, c)}s,$$

$$y = uy_0 - \frac{ay_0}{(a, b, c)}s + \frac{c}{(b, c)}t,$$

$$z = uz_0 - \frac{az_0}{(a, b, c)}s - \frac{b}{(b, c)}t.$$

(g) Find all solutions of each of the following linear diophantine equations.

$$2x + 6y - 8z = 0; \quad 2x + 6y - 8z = 2; \quad 7x - y + 3z = -2.$$

Find all *non-negative* solutions to $2x + 5y + 10z = 100$: the possible combinations of 2p, 5p and 10p which add up to £1. (Or, of course, similar combinations of cents which add up to one dollar!)

Before we can write a program to solve linear diophantine equations, we need a respectable method of finding integers s and t for which $as + bt = (a, b)$. That is provided by the following result.

3.15 Proposition *In the notation of 3.3, define integers s_i and t_i for $i = 0, 1, 2, \ldots$ as follows: $s_0 = 1$, $s_1 = 0$, $t_0 = 0$, $t_1 = 1$ and, for $i \geq 2$,*

$$s_i = s_{i-2} - q_{i-1}s_{i-1}, \quad t_i = t_{i-2} - q_{i-1}t_{i-1}.$$

Define $r_0 = a$, $r_1 = b$. Then $r_i = as_i + bt_i$ for $i \geq 0$, and, in particular, $(a, b) = as_n + bt_n$. (Recall that, in 3.3, r_n is the last nonzero remainder, and is the gcd of a and b.)

Proof As might be expected, the proof is by induction on i and is chiefly a matter of keeping the subscripts under control. Certainly the required formula for r_i is true for $i = 0$ and $i = 1$, by direct substitution. Let $i \geq 2$ and assume that the formula holds for all subscripts j where $0 \leq j \leq i - 1$; we verify it for subscript i. In fact from 3.3,

$r_i = r_{i-2} - r_{i-1}q_{i-1}$

$\quad = as_{i-2} + bt_{i-2} - (as_{i-1} + bt_{i-1})q_{i-1}$, by the induction hypothesis,

$\quad = a(s_{i-2} - s_{i-1}q_{i-1}) + b(t_{i-2} - t_{i-1}q_{i-1})$,

$\quad = as_i + bt_i$, by definition of s_i and t_i. $\hfill \square$

The following program now calculates the successive s_i and t_i, and hence the numbers s and t for which $as + bt = (a, b)$. Note that it is necessary to remember the previous two values of s in order to calculate the current value of s; these are stored as $s1$ (previous value) and $s2$ (one before that) in the program. The same goes for t. You should also ponder on the fact that the value of $q = [a/b]$ at any stage of the program which is being used to calculate the current s and t by the formulas of 3.15 is in fact the current value despite the subscript which q has in 3.15. Look at an example such as $(330, 140)$ if you aren't clear why this is.

3.16 Program for $as + bt = \gcd$ **of** a **and** b, a **and** $b > 0$

```
PROGRAM P1_3_16;
VAR
   a,b,a0,b0,q,r,s,s1,s2,t,t1,t2:   extended;
BEGIN
   writeln ('type two positive numbers separated by a
   space');
   read (a,b);
   a0 := a; b0 := b;
   (* We need to store a and b only to print *)
   (* out the formula as + tb = (a,b) at the end *)
   s2 :=1; s1 := 0; t2 := 0; t1 := 1;
   WHILE b > 0 DO
     BEGIN
       q := INT(a/b);    (* See the note below *)
       r := a - b * q;
       s := s2 - q * s1;
       t := t2 - q * t1;
       (* Here you could print out a0*s + b0*t to make *)
       (* sure it equals r *)
       a := b;
       b := r;
       s2 := s1;   s1 := s;
       t2 := t1;   t1 := t;
       (* Here, the new s-two-steps-back is put equal *)
       (* to the old s-one-step-back, etc., before *)
       (* going on to the next step *)
     END;
   writeln ('the gcd of', a0:0:0, 'and', b0:0:0, 'is',
   a:0:0);
```

```
writeln (a0:0:0, '*', s2:0:0, '+', b0:0:0, '*', t2:0:0,
    '=', a:0:0);
(* You should look carefully at an example to see *)
(* why, at this final stage, it is the s-two-back *)
(* and t-two-back which are used in as + bt = (a,b).   *)
END.
```

Note   As usual we should be a little careful with INT in PASCAL if numbers are < 0. See Note (1) following P1_3_5.

3.17 Computing Exercises

(a) In the case when $(a, b) = 1$, the above program will produce numbers s, t such that $as + bt = 1$, that is, $as - 1$ is a multiple of b. Such an s is called an *inverse of a modulo b*; compare Chapter 3, Section 2. Assume $b > 0$. It is sometimes convenient to add a multiple of b to $s : a(s + mb) + b(t - am) = 1$, in such a way that $0 \le s' = s + mb < b$. Another way of writing this s' is $s - b[s/b]$. Write a program to produce this 'least positive inverse' s'. Beware that in the program the final $s2$ may be negative, and INT rounds towards zero, so you will need lines such as

```
q1 := INT(s2/b);
IF (q1 > s2/b) THEN q1:= q1 - 1;
s2 := s2 - b*q1;
```

Check your program by finding inverses that you can check by hand, and also say: 27 has inverse 104(mod 2807); 19 has inverse 87(mod 1652).

(b) Use 3.16 to write a program for finding a solution of the diophantine equation $ax + by = d$, and hence for finding the general solution. Also, using using 3.14(f), write a program for solving $ax + by + cz = d$. Use the program to solve the equations in 3.14(c) and (g) above.

3.18 Project: $ax + by$ for x and y nonnegative   Let $a > 0, b > 0$

and $(a, b) = 1$. We investigate the set of numbers $A = \{ax + by : x \ge 0 \text{ and } y \ge 0\}$. Show that, if $m \in A$, then $ab - a - b - m \notin A$, and, if $m \notin A$, then $ab - a - b - m \in A$. Here are some hints for the second of these; the first is similar and a little easier. Since $(a, b) = 1$, we can find (by 3.12) integers x, y such that $ax + by = m$; also, for any $k \in \mathbf{Z}$, $a(x + kb) + b(x - ka) = m$; why does this imply that we can choose x such that $0 \le x \le b - 1$? Now suppose (for a contradiction) that m and $ab - a - b - m$ *both* fail to belong to A: then there are integers x, y, u, v such that $ax + by = m$, $0 \le x \le b - 1$, $y < 0$ (else $m \in A$), and $au + bv = ab - a - b - m$, $0 \le u \le b - 1$, $v < 0$. Deduce

$a(x+u+1)+b(y+v+1) = ab$ and use $(a, b) = 1$ to deduce $y+v+1 = \ell a$, $x + u + 1 = (1 - \ell)b$, for some integer ℓ. You are not far now from the required contradiction!

Deduce that all numbers $m > ab - a - b$ are in A and that just half the numbers m with $0 \leq m \leq ab - a - b$ are in A. Write a program to determine, for given coprime a and b, precisely which numbers m do belong to A.

3.19* Exercise   This is based on the result (not the program) of 3.18, so assume that, given a and b (> 0) with $(a, b) = 1$, then all numbers $m > ab - a - b$ can be expressed in the form $ax + by$ for $x \geq 0$ and $y \geq 0$. Suppose you are given a large number of rectangular tiles, size $a \times b$ where a and b are distinct primes (hence certainly coprime), and given a rectangle, $m \times n$, where m and n are both $> ab - a - b$. The idea is to tile the whole of the $m \times n$ rectangle with the $a \times b$ tiles. Of course, this requires $ab \mid mn$. The problem is to prove that, if $ab \mid mn$, then the tiling is possible. Note that, since a and b are primes, the condition $ab \mid mn$ gives essentially two cases: $a \mid m$ and $b \mid n$, or a and b both divide m. Show that in the former case a tiling is possible (this is easy)! Now do the latter case, perhaps starting by choosing w, x, y, z all ≥ 0, such that $aw + bx = m$ and $ay + bz = n$. Of course, you might like to think about smaller values of m and n: is the tiling sometimes possible there too, given $ab \mid mn$? How about values of a and b besides primes?

3.20 Project: number of steps in the Euclidean algorithm
There is a theorem whose precise formulation can be found in [Dixon (1970)]: For 'most' pairs a, b the Euclidean algorithm takes about $(12 \log 2/\pi^2) \log b$ steps. Investigate this numerically, taking random values for a and b.

3.21 Minimal representations and the Euclidean algorithm
(This proof is due to J.W. Bruce.)   Suppose that a and b are positive and coprime and that we use the Euclidean algorithm to find s and t such that $as + bt = 1$. Certainly there are other choices for s and t; in fact the general solution is $a(s + rb) + b(t - ra) = 1$ for any integer r. Let us measure the 'size' of this choice by $f(r) = (s + rb)^2 + (t - ra)^2$. Guided by the hints below you can show that the *minimum* value is always achieved for $r = 0$: the Euclidean algorithm minimizes the 'size' of the coefficients. Now $f(r) = r^2(a^2 + b^2) + r(2sb - 2ta) + (s^2 + t^2)$, a quadratic in r. So the minimum over r is achieved at a single value of r.

(i) To show that the minimum is achieved for $r = 0$ why is it enough to show $f(1) \geq f(0)$ and $f(-1) \geq f(0)$?

(ii) Show that the conditions of (i) hold if and only if $|2sb - 2ta| \le a^2 + b^2$. Thus it is enough to show that when the Euclidean algorithm is used to find s and $t, a \ge |2t|$ and $b \ge |2s|$. This is done by induction on the number of steps in the Euclidean algorithm:

(iii) For one step (so that $b > 1$), $a = bq + 1$, so that $s = 1$ and $t = -q$. Verify the result in this case.

(iv) Suppose that $a = bq + r$ is the first step; then $(b, r) = 1$ and by induction if the algorithm gives $bs_1 + rt_1 = 1$ then $b \ge |2t_1|$ and $r \ge |2s_1|$. Deduce that $s = t_1$ and $t = s_1 - qt_1$. Now deduce that $b \ge |2s|$ and $a \ge |2t|$ as required.

4. $[x]$ and applications

The function 'integer part' was introduced in Section 3 above, and in this section we shall give some properties of the function and a number of applications, mainly to counting the power of a prime which divides certain numbers. Recall the definition:

4.1 Definition   For real x, $[x]$ is the greatest integer $\le x$. Alternative formulations, which clearly give the same value, are:

4.2 *The function $[x]$ is the unique integer satisfying $x - 1 < [x] \le x$.* $[x]$ *is the unique integer with $x = [x] + y$ and $0 \le y < 1$.*

4.3 Exercises

(a) Show that, if n is an integer, then $[x + n] = [x] + n$.

(b) Show that, if n is an integer, then $[x] \ge n \Leftrightarrow x \ge n$.

(c) Show that, if x and y are real, then

$$[x] < [y] \Leftrightarrow \text{ there exists } n \in \mathbf{Z} \text{ with } x < n \le y.$$

(For $\Rightarrow$, n can be chosen as $[y]$.)

(d) Show that $[x + y] = [x] + [y] + \epsilon$, where ϵ is 0 or 1.

(e) Show that, for all real x and y, $[x] + [y] \le [x + y]$. (Of course, this extends to any number of summands.)

(f) Let x and y be ≥ 0. Show that $[xy] \ge [x][y]$. Does this still hold if x or y is < 0?

(g) Show that for any real x, $[x] + [x + \frac{1}{2}] = [2x]$. Here, it is probably easiest to split into two cases, namely $n \le x < n + \frac{1}{2}$ and $n + \frac{1}{2} \le x < n + 1$ for an integer n. Show that, for any real x and integer $n > 0$,

$$\left[\frac{1}{n}[nx]\right] = [x].$$

(h) Let $x \in \mathbf{R}$, and let P and Q be integers with $Q > 0$. Show that there cannot exist an integer k with $[x] + P < Qk \le x + P$. (Compare (c) above.) Deduce that $\left[\frac{[x]+P}{Q}\right] = \left[\frac{x+P}{Q}\right]$. (This result will be used in Chapter 10.) As a special case, note that if a and b are integers with $b > 0$ then

$$\left[\frac{\left[\frac{a}{b}\right]}{b}\right] = \left[\frac{a}{b^2}\right].$$

(i)* Prove the following nice generalization of (h) in [Graham *et al.* (1989), p. 71]. Suppose that f is a continuous, nondecreasing function with the property that, for all x in the domain of f, *if* $f(x)$ *is an integer then* x *is an integer*. Then (assuming continuity on the closed interval $\{y : [x] \le y \le x\}$) we have $[f([x])] = [f(x)]$. (Hints. Let $n = [x]$ and assume $n < x$. Show that $[f(n)] \le [f(x)]$ and assume $<$ for a contradiction, so that $[f(n)] + 1 \le [f(x)]$. Deduce that $f(n) < [f(x)] \le f(x)$ and use continuity to show that there exists y with $n < y \le x$ and $f(y) = [f(x)]$. Since y must be an integer (from the assumptions about f) this is a contradiction.)

(j) Let r and i be integers with $i > 0$. Let $x = \frac{r+i}{2i}$. Use $[x] \le x$ to show that $2[x] - 1 < \frac{r+1}{i}$, and use $x - 1 < [x]$ to show that $r - i \le 2i[x] - 1$. Deduce that $\frac{r+1}{i} \le 2[x] + 1$ and $2[x] - 1 < \frac{r+1}{i}$, and hence that the smallest *odd* integer $\ge \frac{r+1}{i}$ is $2\left[\frac{r+i}{2i}\right] + 1$. (This will be used in the next chapter; see Chapter 2, Section 2.) What is the condition for the above odd number to be 1 ?

(k) Let $x > 0$. Show that the number of odd numbers ≥ 1 and $\le x$ is $[(x+1)/2]$.

(l) Let k be an integer ≥ 2. Divide 3^k by 2^k: $3^k = q \times 2^k + r$, where $0 \le r < 2^k$ so that $q = [3^k/2^k]$. Let $s = q \cdot 2^k - 1$, which is $< 3^k$. Suppose that s is expressed as a sum of kth powers of (not necessarily distinct) positive integers. Why can these only be k^{th} powers of 1 or 2? Now use $s = (q-1) \cdot 2^k + (2^k - 1) \cdot 1^k$ to deduce that s is a sum of $(q - 2 + 2^k)$ k^{th} powers and no fewer. Thus there exist integers which can be expressed as a sum of $[3^k/2^k] - 2 + 2^k$ k^{th} powers and no fewer. What does this give for $k = 2, 3, 4$? [This is related to *Waring's problem*; see 5.8 below.]

(m) Show that, for n a positive integer,

$$(n + \frac{2}{5})^2 < n(n+1) < (n + \frac{1}{2})^2, \quad (n + \frac{7}{10})^2 < n(n+2) < (n+1)^2$$

and

$$(n + \frac{7}{5})^2 < (n+1)(n+2) < (n + \frac{3}{2})^2.$$

Deduce that $[\sqrt{n} + \sqrt{n+1} + \sqrt{n+2}] = [\sqrt{9n + 8}]$.

(n) Let n be an integer ≥ 1 and $s = [\sqrt{n}]$. Show that $s^2 \le n \le s^2 + 2s$

and that

$$[\tfrac{n}{s+1}] = \begin{cases} s-1 & \text{if } s^2 \le n < s^2 + s, \\ s & \text{if } s^2 + s \le n \le s^2 + 2s. \end{cases}$$

$$[\tfrac{n}{s}] = \begin{cases} s & \text{if } s^2 \le n < s^2 + s, \\ s+1 & \text{if } s^2 + s \le n < s^2 + 2s, \\ s+2 & \text{if } s^2 + 2s = n. \end{cases}$$

In particular show that $[n/s] = s + 2$ if and only if $n + 1$ is a perfect square.

(o) Suppose that $i^2 \le n (i > 0)$. Show that

$$\left[\frac{n}{[n/i]}\right] = i.$$

(One way is to write $n = ai + b$ where $0 \le b < i$, in which case you want to show $a > b$. Assuming $a \le b$ deduce $i \ge a + 1$, and then using $n < (a + 1)i$ derive a contradiction.) Show further that if $[n/i] = u$ and $[n/(i+1)] = v$ (and still supposing $i^2 \le n$), then $[n/w] = i$ for all w with $v + 1 \le w \le u$.

4.4 The power of a prime dividing $n!$

Let $n \ge 2$ and let p be prime. We want to find the exact power of p which divides $n!$. For example, if $n = 6$, then $n! = 2 \cdot 3 \cdot 4 \cdot 5 \cdot 6 = 2^4 \cdot 3^2 \cdot 5$ and for $p = 2$ the answer would be 4, for $p = 3$ it would be 2 and for $p = 5$ it would be 1, while for any other prime the answer would be 0. Notice that, for $p = 2$ say, the factors among $2, 3, 4, 5, 6$ which are divisible by 2 are just $2, 4, 6$. But we have to count 4 an extra time since it is divisible by 2^2 and not just 2. So the total power is $3 + 1, 3$ coming from $2, 4, 6$ and 1 extra coming from 4.

The general principle is the same for any p and n. Since $n! = 2 \cdot 3 \cdot 4 \cdot \ldots \cdot n$ and p is *prime*, we want to count the total number of times p goes into the factors $2, 3, 4, \ldots, n$ of $n!$ To do this we start by counting the number of multiples of p among $2, 3, 4, \ldots, n$. We then count the multiples of p^2, each of which gives an *additional* factor of p in $n!$. We then count the multiples of p^3, each of which gives a factor of p in $n!$ additional to those counted so far, and so on. The multiples of any integer $k \ge 2$ among $2, 3, \ldots, n$ are just $k, 2k, 3k, \ldots, [n/k]k$, hence are $[n/k]$ in number. Thus the number of multiples of p is $[n/p]$, of p^2 is $[n/p^2]$, etc. The end result is the following:

The power of a prime p which divides $n!$ is $\left[\frac{n}{p}\right] + \left[\frac{n}{p^2}\right] + \left[\frac{n}{p^3}\right] + \ldots,$ where the sum is continued until the terms become 0, that is, until $p^r > n$.

The reader may like to prove the following equivalent form of this result: the power of p dividing $n!$ is $(n-s)/(p-1)$, where s is the sum of the digits of n when expressed in base p. (That is, if $n = \sum_{i=0}^{t} a_i p^i$, where $0 \le a_i < p$ for all i, then $s = \sum_{i=0}^{t} a_i$.)

For example, let $n = 100$ and $p = 2$. The power of 2 dividing $100!$ is then $\left[\frac{100}{2}\right] + \left[\frac{100}{4}\right] + \left[\frac{100}{8}\right] + \left[\frac{100}{16}\right] + \left[\frac{100}{32}\right] + \left[\frac{100}{64}\right] = 50+25+12+6+3+1 = 97$. Similarly if $n = 100$ and $p = 5$ then the power of 5 dividing $100!$ is $\left[\frac{100}{5}\right] + \left[\frac{100}{25}\right] = 20 + 4 = 24$. Thus $100! = 2^{97} \cdot 5^{24}$ times other prime powers (including a power of 3, of course). In particular the largest power of 10 which divides $100!$ is 24, the smaller of 97 and 24: there are 24 zeroes at the end of $100!$.

4.5 Computing exercises

(a) Write a (short!) program to work out the numbers $([\sqrt{n}]+1)^2 - n$ for $n = 1, 2, 3, \ldots, 100$. Now explain what the program produces. (Hint: If $(m-1)^2 \le n < m^2$, what is $[\sqrt{n}]$?)

(b) Write a program which uses the result of 4.4 to calculate the power of a prime p which divides $n!$, for given p and n. Use it also to calculate the power of p which divides the binomial coefficient

$$\binom{n}{k} = \frac{n!}{k!(n-k)!}.$$

Find the prime-power factorization of $50!$ and of $\binom{50}{17}$.

4.6 Project
There is another way to calculate the power of p which divides $\binom{n}{k}$: it is the number of 'carries' in the addition of k and $n-k$ in base p. This result was originally due to E.E. Kummer in the 19th century, but was rediscovered in [Goetgheluck (1987)], where it is also explained in detail. Write a program to implement it for given n, k and p. There are also generalizations to 'multinomial coefficients' in [Dodd and Peele (1991)].

4.7 Project: sets of powers
For a fixed n and a given prime p, consider the set $E(n, p)$ of highest powers of p which divide the various binomial coefficients $\binom{n}{k}$ for $k = 0, 1, \ldots, n$. For example, with $n = 10$, the distinct binomial coefficients are 1, $2 \cdot 5$, $3^2 \cdot 5$, $2^3 \cdot 3 \cdot 5$, $2 \cdot 3 \cdot 5 \cdot 7$, $2^2 \cdot 3^2 \cdot 7$. Thus $E(10, 2) = \{0, 1, 2, 3\}$ and $E(10, 3) = \{0, 1, 2\}$. Examine some values of n up to say 50; do you notice anything about the sets $E(n, p)$? You can read about this in [Wong 1989].

4.8 Exercises
(a) Why does the formula of 4.4 not give the correct answer when p is not a prime?

(b) Find the number of zeroes at the end of 1000! and at the end of the binomial coefficient $\begin{pmatrix} 1000 \\ 353 \end{pmatrix}$.

(c) Using the result of 4.3(e), show that the largest power of a prime p which divides $(a+b+c)!$ is greater than the largest power which divides $a!\,b!\,c!$ (for integers $a,b,c \geq 1$). Deduce that the 'trinomial coefficient'

$$\frac{(a+b+c)!}{a!\,b!\,c!}$$

is an integer.

(d) Show that, if p is prime and p^r divides the binomial coefficient $\begin{pmatrix} n \\ k \end{pmatrix}$, then $p^r \leq n$. [Hint. The power of p dividing the binomial coefficient is a sum of terms of which the first is $\left[\frac{n}{p}\right] - \left[\frac{k}{p}\right] - \left[\frac{n-k}{p}\right]$. Use 4.3(d) to show that this is 0 or 1.]

(e) Let n be composite, $n = p_1^{k_1} p_2^{k_2} \ldots p_r^{k_r}$, where $r \geq 2$. Why must each prime power $p_i^{k_i}$ be one of the numbers $1, 2, \ldots, n-1$? Deduce that this prime power divides $(n-1)!$ and hence that $n \mid (n-1)!$.

Now let n be composite but $r = 1$ above so n is a prime power. Assume that $n > 4$. Write down the formula for the power of p dividing $(n-1)!$ and show that this power is $\geq k$ (it is enough for the first term of the sum in 4.4 to be ≥ 2 and all the other terms to be ≥ 1). Deduce that $n \mid (n-1)!$ in this case too.

(f) Let s be odd and > 1, $2^t < s < 2^{t+1}$. Show that

$$\frac{s}{2} + \frac{s}{4} + \ldots + \frac{s}{2^t} < s - 1,$$

and deduce that

$$\left[\frac{s}{2}\right] + \left[\frac{s}{4}\right] + \left[\frac{s}{8}\right] + \ldots < s - 1,$$

where the sum is continued until the terms become zero.

Now let $n = 2^r s$, where s is odd and > 1, and $r \geq 1$. Show that the power of 2 dividing $n!$ is $< n - 1$. Deduce that $2^{n-1} \mid n!$ if and only if n is a power of 2.

Show similarly that the highest power of p dividing $r!$ is $< r/(p-1)$ (so certainly $\leq r - 1$).

(g) Suppose that p is prime and $p^a \mid (m-1)$. Show that $p^{a+b} \mid (m^{p^b} - 1)$ for any $b \geq 0$.

(h) Show that

$$\frac{(2a)!\,(2b)!}{a!\,b!\,(a+b)!}$$

is always an integer.

(i) Let $n > 1$. Why is the power of 2 which divides $n!$ always greater than the power of 5 which divides $n!$? Deduce that the last nonzero decimal digit of $n!$ (reading left to right) is always even. Show also that the number of zeroes at the end of $(5n)!$ is n more than the number of zeroes at the end of $n!$. (Harder: Writing $f(n)$ for the last nonzero digit of $n!$ show that $f(5n)$ is the last digit of $2^n f(n)$.)

4.9 Project: the function $S(n)$

This function is defined by

$$S(n) = \sum_{d=1}^{n} \left[\frac{n}{d}\right].$$

Write a program to calculate this directly from the definition and observe the running times of the program for values of n around $10\,000$.

Now let $s = [\sqrt{n}]$ and $q_i = [n/i]$ for $i = 1,2,\ldots,s$. Consider the sum of $[n/d]$ for d from q_s to n. By breaking this sum up into $[n/q_s]$ + (those terms ending in $[n/q_{s-1}]$) + (those terms ending in $[n/q_{s-2}]$) + and so on, show that the sum is

$$q_1 + q_2 + \ldots + q_s + \left[\frac{n}{q_s}\right] - sq_s.$$

Deduce that $S(n) = -s^2 + 2\sum_{d=1}^{s} \left[\frac{n}{d}\right]$ (You may find 4.3(n) and (o) helpful.) Use this to calculate the exact values of $S(n)$ for $n = 10^k$ where $k = 5, 6, 7, 8$. Estimate the value of the limit $\lim_{n\to\infty} \left(\frac{S(n)}{n} - \log(n)\right)$. For further details and another spectacular simplification of a large sum, see [Hoehn and Ridenhour (1989)].

5*. Further projects.

We conclude this chapter with a collection of fairly challenging projects which do not involve ideas from the remainder of the book (although later work may shed additional light on them). Several of them involve only the most elementary ideas from number theory, but they require a certain ingenuity to turn into computer programs. We hope you enjoy trying them.

5.1 Kaprekar constants Let b be a positive integer (the base) and let a be a positive integer with a four-digit expansion in base b (e.g. $b = 10$, $a = 2674$), not all digits being the same. Define $T(a) = a' - a''$, where a' (resp. a'') is the integer written in base b obtained by writing the digits of a in descending (resp. ascending) order. (e.g. $b = 10$, $a = 2674$, $a' = 7642$, $a'' = 2467$, $T(a) = 5175$.) For $b = 10$ show that $T(a) = a$ if and only if $a = 6174$. Start by considering a general four-digit number *pqrs* (these are the decimal digits) with $p \geq q > r \geq s$, and show that subtracting *srqp* gives a difference whose outer digits sum to 10 and whose middle digits sum to 8. A similar result holds when $q = r$. This means that the possibilities for $T(a)$ are very restricted.

The number 6174 is called *Kaprekar's constant* for base 10. For $b = 5$ find the unique a for which $T(a) = a$: this is called Kaprekar's constant for base 5. Show that, in both cases, for any starting value of a, the sequence $T(a), T^2(a) = T(T(a)), T^3(a) = T(T^2(a)), \ldots$ reaches the Kaprekar constant for that base. (For $b = 10$, consider the restricted range of numbers which can equal $T(a)$, found above.)

Write a program to iterate the operation T for any chosen b (perhaps restricting to $b \leq 10$) and hence to find the Kaprekar constant for base b when it exists. References: [Kaprekar (1955), Jordan (1964), Eldridge and Sagong (1988), Lines (1986), pp. 53-61.]

5.2 Hilbert primes   Let $H = \{4k + 1 : k = 0, 1, 2, 3, \ldots\}$. Show that H is closed under multiplication (i.e., if $a \in H$ and $b \in H$ then $ab \in H$). An integer $h \in H$ is called a *Hilbert prime* if, whenever $h = ab$ for a and $b \in H$, then $a = 1$ or $b = 1$. Write a program to find Hilbert primes. Find integers which factorize in *two different ways* as products of Hilbert primes. (So the fundamental theorem of arithmetic, 1.1, which asserts that for ordinary primes the factorization is *unique*, really does say something substantial after all!)

5.3 Farey series   Write a program to find the *Farey fractions of order n,* that is, the fractions p/q in lowest terms (i.e., p and q coprime), such that the denominator q is $\leq n$ and (here's the catch) the fractions are arranged in increasing order of magnitude. See for example [Rademacher and Goldfeld (1983), Chapter 4.]. There is an algorithm for passing from Farey fractions of order n to those of order $n + 1$: between each consecutive pair of the former, $a/b < c/d$ say, insert the 'mediant' $(a + c)/(b + d)$, and then discard any fractions for which $b + d > n + 1$.

5.4 Fibonacci nim   Write a program to express any number m in the 'Fibonacci system', that is, as a sum of distinct and nonadjacent Fibonacci numbers.   These numbers (named after Fibonacci, alias Leonardo of Pisa (1180–1228)) are defined by $f_1 = 1$, $f_2 = 1$, $f_n = f_{n-1} + f_{n-2}$ for $n \geq 3$. Thus the sequence starts out 1, 1, 2, 3, 5, 8, 13, 21, 34, 55, ... As an example, 54 can be written as a sum $34 + 13 + 5 + 2$: these are all Fibonacci numbers, and no two come together in the sequence of f_i. The idea is to subtract from m the largest $f_i \leq m$, and then from $m - f_i$ the largest $f_j \leq m - f_i$, and so on.

Now consider the following game played by two players who take chips alternately from one pile of n chips. The first player is allowed to take any number $m_1 < n$ of chips; the second can take any number m_2 where $0 < m_2 \leq 2m_1$. From then on the players alternate, each one taking at least one chip and not more than twice the preceding 'grab'. The last grabber wins.

The strategy is as follows. If the number of chips at any stage is not a Fibonacci number then it is expressed as a sum of Fibonacci numbers as above, e.g. $54 = 34 + 13 + 5 + 2$. The next player to move then takes any number of small terms from the expansion, provided their total is less than half the next larger term. Thus from a pile of 54 this player would take only 2, not $2 + 5 = 7$, since $2 \times 7 > 13$, and the other player could take 13 and leave a Fibonacci number of chips, namely 34.

Write a program to play Fibonacci nim, Computer versus Player. If you can explain the strategy, so much the better! [See Berlekamp *et al.* (1982), Vol. 2, p. 483, Anderson (1991)]

5.5 Palindromes by reversal   Example: $87 + 78 = 165$, $165 + 561 = 726$, $726 + 627 = 1353$, $1353 + 3531 = 4884$, which is a palindrome, i.e., reads the same backwards as forwards. Write a program to take numbers up to say 999 and repeat the above operation until either the number is a palindrome or it is too large. Many numbers form palindromes very quickly; try 187 as starting number for an example where this does not happen.

Make a table of results in the form of the number of steps taken for each starting number. (According to [Wells (1986), entry 196] only 75 of the 900 three-digit numbers require more than five reversals, and all numbers $< 10\,000$ have produced a palindrome eventually with the exception of 196, which hasn't yet produced one! See also [Gardner (1970) and Lines (1986), pp. 62–65].)

It is interesting to note that binary numbers do not always give palindromes. Here is a suggestion for proving that 10110 never gives one. Show that after four iterations it has become 10110100, or say 101_2010_2 where the suffices indicate the number of repeated digits. Show that four further iterations produce 101_3010_3, and so on, and deduce that from the second iteration onwards the numbers alternate between $10\ldots00$ and $11\ldots01$ where the $\ldots$ indicates some sequence of 0s and 1s.

5.6 Shunting goods wagons   This problem concerns shunting goods wagons on to multiple sidings. The reference is [Schroeder (1986), p. 75]. We define an *index* $m(\sigma)$ of a sequence σ of integers such as

5632413572623 (generally $a_1, \ldots, a_r$ with $1 \le a_i \le n$ for each i, and every integer from 1 to n inclusive occurring at least once).

(i) Start at the left and write down all the 1s, then all 2s (if any) to the right of all these 1s, then, *provided all the 2s in the sequence have now been used up,* all 3s (if any) to the right of all these 2s, etc. In the example we get 122. Now delete these from the sequence.

(ii) Start at the left of the new (shorter) sequence and write down all occurrences of the smallest integer a_i, then all $a_i + 1$ (if any) to the right of all these a_i, then, *provided all $a_i + 1$s have now been used up,* all $a_i + 2$s to the right of all these, etc. In the example this gives 233.

(iii) Repeat until the sequence is exhausted. In the example, this gives 345, then 566 and finally 7.

Then $m(\sigma)$ is the number of these nondecreasing sub-sequences you end up with. In the example, $m(\sigma) = 5$, because the sub-sequences are 122, 233, 345, 566, 7.

Write a program to calculate $m(\sigma)$. The connexion with shunting goods wagons is explained in [Schroeder (1986), §5.8]. Roughly speaking, it is the number of shuntings needed to change the initial order $56324\ldots$ of marked goods wagons into the order 1222333455667, using one engine and a number of sidings.

5.7 The $3N + 1$ problem This is a well-known but very baffling iteration procedure. Start with any positive integer N and apply the following function to N:

$$f(N) = \begin{cases} 3N + 1 & \text{if } N \text{ is odd,} \\ N/2 & \text{if } N \text{ is even.} \end{cases}$$

The question is : does the sequence $N, f(N), f(f(N)), \ldots$ eventually become 1 (and therefore go through the cycle $1, 4, 2, 1, \ldots$ thereafter)? It is of course extremely easy to write a program to iterate f; what is harder is to make any sense of the resulting sequences of numbers for different starting values. For instance, starting with $N = 27$, the sequence reaches several peaks, the highest of which is 9232 at the 77th iteration, before it ends in 1 at the 111th iteration. The maximum value 9232 also occurs for several other values of N such as $54, 73$ and 97, which have the property that their 'path lengths' (the number of iterations to reach 1) is greater than that of all smaller starting numbers N. Try to find these values of $N (< 700)$.

You can find an introduction to this weird problem in [Hayes 1984]; see also [Lagarias (1985)] for a wealth of information and many references. Note that Lagarias uses the function in which $3N + 1$ is replaced by

$(3N + 1)/2$. This is essentially equivalent since, for odd N, $3N + 1$ is even, so the next iteration takes it to $(3N + 1)/2$.

5.8 Waring's problem It is a famous theorem of J.L. Lagrange (1736–1813) that every positive integer n can be written as the sum of at most four squares. For example, $7 = 2^2+1^2+1^2+1^2$, $15 = 3^2+2^2+1^2+1^2$ cannot be expressed as a sum of fewer than four squares but $8 = 2^2 + 2^2$ and $11 = 3^2 + 1^2 + 1^2$ can. (Many books on number theory contain an account of this theorem, for example [K.H. Rosen (1988), §11.3; Ribenboim (1988), pp. 236ff].) For sums of positive *cubes* the situation is not quite so well understood. It is known that nine cubes always suffice, and that indeed seven suffice for all sufficiently large n (it is conjectured that in fact four cubes suffice for all sufficiently large n). Write a program to determine, for all $n \leq 10\,000$ say, what is the smallest number of positive cubes whose sum equals n. In particular check that those which need all nine are precisely 23 and 239 (these are in fact the *only* two numbers which need nine cubes), and list those that need eight, and those that need seven.

Here is a suggestion for making the program reasonably efficient. Set up an array, say d, by the declaration

    d: array [0..10000] of integer;

and define `d[0] := 0`, `d[1] := 1`. The number $d[j]$ is to be (eventually) the smallest number of cubes required to express j. For each $j = 2, 3, \ldots, 10\,000$ in turn the value of $d[j]$ starts at $d[j-1]+1$ (corresponding to writing $j = 1^3 +$ (expression for $j-1$ as a sum of cubes)). Now consider all numbers of the form $t = j - i^3 \geq 0$ (where $i \geq 2$); if $1 + d[t] < d[j]$ then replace $d[j]$ by $1 + d[t]$. (Note $d[t]$ has already been determined since $t < j$.) Do you see why it is necessary to define $d[0] = 0$? (Try for example $j = 8$.) Perhaps you can devise a program which actually produces the cubes required for a given j.

This is a special case of *Waring's problem* (stated by E. Waring in the form of a conjecture in 1770): for a fixed $k \geq 2$, determine the smallest number of positive k^{th} powers needed to express (a) every integer, (b) every sufficiently large integer. For $k = 3$ the answer to (a) is 9 but the answer to (b) is 4, 5, 6 or 7. A very simple argument which suggests the answer to (a) is given in 4.3(1) above. See also Chapter 3, 1.4(c).

5.9 Zigzags (The idea for this project comes from [Abelson and diSessa (1980), p. 114]. This book is a wonderful source of ideas on the interaction between mathematics and graphics.)

Consider the following construction. Start with a horizontal line of length 100 units, and give yourself two angles α and β, which we shall

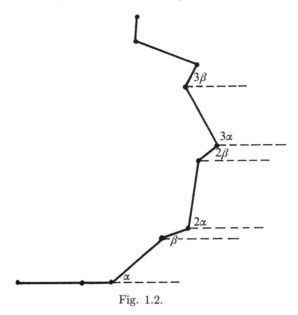

Fig. 1.2.

assume to be nonzero whole numbers of degrees, and a length $\ell \geq 0$. (This can be any real number, though in practice it might as well be an integer.)

At the end of the line draw another horizontal line of length ℓ. Now we start to zigzag by drawing a sequence of lines of lengths 100, ℓ, 100, ℓ, 100, ℓ, ..., the anticlockwise angles these lines make with the horizontal being α, β, 2α, 2β, 3α, 3β, ... (see Fig. 1.2). We call all the lines of length 100 (including the first horizontal line), zigs, and similarly all the lines of length ℓ, zags.

The interesting thing is that the zigzags always close up. Some very pretty patterns are obtained by judicious choices of ℓ, α and β. Write a graphics program to draw zigzags. It is a good idea to have the zigs drawn in one colour and the zags in another. Here are some triples to try: $(100, 90, -40)$, $(100, 62, -60)$, $(100, 19, -20)$, $(40, 45, 9)$, $(57, 32, 4)$, $(54, 35, 79)$, $(200, 5, 6)$, $(200, 299, 45)$, $(100, 175, 185)$. You should also try making some simple shapes such as a regular hexagon, a circle (or a regular polygon with a very large number of sides!), and, harder, an ellipse.

The number of steps (one step = one zig and one zag) which are needed to close the zigzag is given by the formula $360/(360, \alpha, \beta)$; thus if α, β and 360 have no common factor then the number of steps is 360. Here is a suggestion for proving this formula. You can incorporate a calculation of the number of steps into your program and stop it after that number of steps. Let u and v be vectors along the first zig and zag, respectively (so in fact u and v are horizontal, of lengths 100 and ℓ, respectively). Write ρ for rotation through α and σ for rotation through β. (Note that rotation of vectors is well-defined without specifying a centre of rotation.) Let the initial point of the zigzag be the origin. Then after k steps the end of the last zag is at the point

$$u + v + \rho u + \sigma v + \rho^2 u + \sigma^2 v + \ldots + \rho^{k-1} u + \sigma^{k-1} v$$

$$= (u + \rho u + \ldots + \rho^{k-1} u) + (v + \sigma v + \ldots + \sigma^{k-1} v).$$

For this to be the origin it is certainly *sufficient* for each bracket to be zero separately. (In fact if, in addition, the next zig and zag after returning to the origin are to be along the first zig and zag, then it is also *necessary* that each bracket be zero.) Putting the first bracket equal to 0 and operating on it by ρ, show that $u = \rho^k u$, which implies ρ^k is the identity rotation (we assume $u \neq 0$). This requires that $k\alpha$ is a multiple of 360; why is this the same as k a multiple of $360/(360, \alpha)$? When both brackets are zero, k must be a common multiple of this and of $360/(360, \beta)$.

A worthwhile refinement of the program is to estimate the size of the completed zigzag and then scale the drawing to fit on the screen. Here, without proof, is one solution to this; you may like to find your own. Arrange for the centre of the screen to have coordinates $(50, 50 \cot(\alpha/2))$, and for the height of the screen to be $|50 \mathrm{cosec}(\alpha/2)| + |(\ell/2)\mathrm{cosec}(\beta/2)|$ units. (Recall that the starting point of the zigzag is $(0,0)$ and the first zig is horizontal of length 100.)

5.10 Project: Fibonacci numbers   These numbers $f_1, f_2, f_3, \ldots$ are defined in 5.4. An extraordinary theorem (see [Ribenboim (1988), p. 149 and Jones (1975)]) states that the f_i are precisely the positive values of the polynomial

$$2xy^4 + x^2 y^3 - 2x^3 y^2 - y^5 - x^4 y + 2y,$$

where x and y are integers ≥ 0. Such pairs (x, y), where $x + y = n$ say, can be enumerated as $(0, n)$, $(1, n-1)$, $(2, n-2)$, $\ldots$, $(n, 0)$. By doing this for $n = 1, 2, 3, \ldots$ in succession verify with a suitable program that the theorem holds up to say $n = 50$. Can you conjecture the values of x and y which produce a given f_i as the value of the polynomial?

5.11 Project: verbal sequences   Consider sequences of the form $s_1 = 40$, $s_2 = 14/10$, $s_3 = 14/21/10$, $s_4 = 14/12/31/10, \ldots$. The interpretation of these is verbal: in s_1 there is one four and one zero;

this is recorded in s_2. In s_2 there is one four, two ones and one zero; this is recorded in s_3. In s_3 there is one four, one two, three ones and one zero; this is recorded in s_4, and so on. Verify that the above sequence eventually becomes periodic with period 2. Write a program to calculate such sequences from any starting term. Only digits 0, 1, 2, ..., 9 are involved in the counting process and you may if you wish restrict to sequences each term of which has the form $a_9 9/a_8 8/\ldots/a_0 0$ where $0 \le a_i \le 9$ for every i (that is, abandon any sequence which produces a term violating this condition). Investigate the existence of cycles where $s_{n+k} = s_n$ for some n and k (e.g. $k = 1$ gives a 'fixed point' where the sequence remains constant). There is a full discussion of this idea in [Lehning (1990)].

5.12 Project: using all the numbers 1 to 9   If we take a five-digit number and a four-digit number, with all nine digits different and > 0, then occasionally the quotient is a whole number. For example,

$$\frac{13458}{6729} = 2; \quad \frac{98736}{1452} = 68.$$

Write a program to find all such pairs of numbers (there are 187 altogether!). [This can be done fairly efficiently with nine nested loops in the program, one for each digit. You will probably want to use an array of 9 entries to keep track of which digits have been used. The natural way to print out the solutions is in increasing order of the number formed by all nine digits.]

2

Listing primes

In this chapter we shall give three methods of listing primes in ascending order: by trial division, by multiplication to eliminate composites, and by sieving. At least one of these three methods should be used to produce a text file containing the primes up to about 100 000. This will be very useful in later computing work.

1. Primes by division

We begin with a very useful and fundamental fact.

1.1 Proposition *Let n be composite, $n > 1$. Then n has a prime factor p satisfying $p \leq \sqrt{n}$. Equivalently: if $n > 1$ has no prime factor $p \leq \sqrt{n}$, then n is prime.*

Proof Let $n > 1$ be composite, $n = ab$ where $a > 1$, $b > 1$. Then either a or b must be $\leq \sqrt{n}$, since if both were $> \sqrt{n}$ then ab would be $> n$. Suppose $1 < a \leq \sqrt{n}$. Then a has a prime factor p, $p \leq a$, and consequently $p \leq \sqrt{n}$. $\qquad\square$

1.2 Exercises
(a) Which numbers n have the property that the smallest prime factor of n is *equal* to $\sqrt{n}$?

(b) Let n be composite. Show that, if the smallest prime factor p of n is greater than the cube root of n, then n/p is prime.

(c) Let p be prime, and let $p = 30q + r$, $0 \leq r < 30$. Show that r is either 1 or a prime. (Why is it enough to prove that, unless $r = 2, 3$ or 5, r is not divisible by 2, 3 or 5?) In fact 30 is the *largest* number for which this works; can you find other *smaller* numbers for which it works?

We now want to list the primes by checking numbers for divisibility by primes up to their square root. There are at least two approaches to this, one rather more sophisticated than the other. Let us for a start confine attention to *odd* numbers. All even numbers besides 2 are composite so they can be dismissed at once. The first, more straightforward, method takes each candidate n in turn (that is, $n = 3, 5, 7, 9, 11, \ldots$) and trial-divides by *all odd numbers* $\leq \sqrt{n}$. Clearly if no factor is found then no *prime* less than $\sqrt{n}$ can divide n, so n is prime. The following program performs this trial division for all numbers $n \leq$ max, printing out those which are prime.

1.3 Program for primes by trial division by odd numbers up to the square root

```
PROGRAM P2_1_3;
VAR
n,q,max:extended;
qdividesn:boolean;

BEGIN
write(' Max = ?  ');
    (* Prompts you for the maximum:  *)
    (* all primes ≤ max are printed out *)
readln(max);
write('2');
n:=3;
WHILE n<=max DO
  BEGIN
  q:=3;
  qdividesn:=false;
  (* Initialization for the WHILE loop *)
  WHILE (q*q<=n) AND NOT(qdividesn) DO
    BEGIN
    qdividesn:=(n=q*INT(n/q));
    q:=q+2;
    END;
  IF NOT(qdividesn) THEN write(',',n:0:0);
  n:=n+2;
```

```
    END;
END.
```

There is a slightly more sophisticated way of doing the same thing
as 1.3, but trial-dividing only by those *primes* $\le \sqrt{n}$, rather than by
all odd numbers $\le \sqrt{n}$. This assumes that primes $\le \sqrt{n}$ are already
available, and it necessitates storing the primes found in an array so
that they continue to be available during the running of the program.
Rather than giving a program which does exactly the same as 1.3, here
is a program, using the above idea, which prints out a given number of
primes, referred to as no_primes in the program. The primes are stored
in an array $p[1..no\_primes]$.

**1.4 Program for primes by trial division by primes up to the
square root**

```
PROGRAM P2_1_4;
CONST
size = 1000;
(* This is the maximum number of primes to be found, *)
(* and can be adjusted as required *)
VAR
n:  extended;
i,j,no_primes:  integer;
p:  array[1..size] of extended;
pass:  boolean;

BEGIN
write ('Number of primes = ?   ')
readln (no_primes);   (* This is ≤ size *)
p[1]:= 2;
p[2]:= 3;
writeln(2,',',3); n:= 3;
FOR i:= 3 TO no_primes DO
  BEGIN
  pass:= false;
  WHILE NOT(pass) DO
  (* So long as a new prime is not *)
  (* discovered the program *)
  (* adds 2 to n and tries again *)
    BEGIN
    n:= n + 2;
    j:= 2;
```

```
(* Start by trial-dividing by the 2nd prime, 3 *)
pass:= true;
WHILE (pass AND (j < i) AND (p[j]*p[j]<=n)) DO
  BEGIN
  pass:= NOT(n = p[j]*INT(n/p[j]));
  (* The number n passes this test if it is not *)
  (* divisible by the jth prime *)
  j:= j + 1;
  END;
IF pass THEN
(* If n passes all the tests it is added to p *)
  BEGIN
  p[i]:= n;
  write(',',n:0:0);
    (* and also written to the screen *)
  END;
END;
(* If NOT(pass) then 2 is added to n and we
(try again; *)
(* if pass then we go on to the next value of i *)
END;
END.
```

1.5 Computing exercise   Modify the program 1.3 to count the number of primes up to the limit max. Thus a counting variable (of integer type) needs to be introduced and incremented by one every time a new prime is found. Let us write $\pi(x)$ for the number of primes $\leq x$. Use the program to verify the following:

x	$\pi(x)$		x	$\pi(x)$		x	$\pi(x)$
1000	168		4000	550		7000	900
2000	303		5000	669		8000	1007
3000	430		6000	783		9000	1117

How would you verify these figures using 1.4 instead of 1.3?

Before going on to produce a file of primes, and to write a program for factorization by trial division, let us investigate briefly the gaps between successive primes. (Note that these gaps *can* be arbitrarily large; compare Chapter 1, 1.5(e).) Given an integer $k \geq 2$, we want to find those consecutive primes below the maximum under consideration which differ by k. In the special case $k = 2$ these pairs of primes are called *twin primes.* Examples of twin primes are 3, 5; 5, 7; 11, 13; 17, 19. It was conjectured a long time ago that there are infinitely many twin primes, but this has not been proved. You can find some information on twin primes in, for example, [Ribenboim (1988), pp. 199ff]. Ribenboim lists some gigantic twin primes, such as the 1040-digit primes

$$256\,200\,945 \times 2^{3426} \pm 1.$$

These are somewhat beyond our reach here, but you can investigate more modestly sized twin primes for yourself and a project on them is suggested in 1.7 below.

It is very easy to modify the program 1.4 to print out successive primes which differ by k: you need only check whether $p(i) - p(i - 1) = k$ for each i in turn. It is marginally more difficult to do the same job with program 1.3, since the primes are not stored as they are found. The trick is temporarily to store the value of each prime as it is found, under a name such as *oldn*, initializing *oldn* to 3 at the beginning of the program (and declaring *oldn* and k even earlier!) and then checking as each new prime is found whether $n - oldn = k$.

1.6 Computing exercise: prime gaps   Use the above idea to adapt program 1.3 so that it prints out the primes which differ from the previous prime by k, where k is input to the program by the user. Find the first two consecutive primes which are (a) 10 apart, (b) 20 apart. What happens when $k = 25$? There is a lot of information about gaps between primes in [Young and Potler (1989)].

1.7 Computing exercise: twin primes   Modify 1.6 to use only $k = 2$ and to count the number of twin primes (starting with 3,5) below a given maximum. Verify that there are 35 (pairs of) twin primes < 1000 and 205 twin primes $< 10\,000$. (It is conjectured that the number of twin primes $\leq N$ is, for large N, given by $\alpha \int_2^N \frac{dx}{(\log^2 x)}$, where α is a constant whose approximate value is 1.320 323 6. If you know about approximating integrals by Simpson's rule, you could test this conjecture.)

1.8 Computing exercise: prime triples   We can't expect to find three primes of the form p, $p + 2$, $p + 4$ apart from 3, 5, 7 (why not?) but there are plenty of examples of prime triples of the form $6k - 1$, $6k + 1$, $6k + 5$. Adapt 1.6 to print out all such prime triples which are less than some maximum. It is conjectured that infinitely many such triples exist (see [Guy (1981), §A9]). Another possibility is to look for *prime quadruplets* p, $p + 2$, $p + 6$, $p + 8$. The first two are 5, 7, 11, 13 and 11, 13, 17, 19.

1.9 Computing exercise: largest gaps between primes   Choosing a maximum, *max*, adapt 1.3 to find the largest gap which occurs between consecutive primes both of which are ≤ *max*. Now you need to keep a record of the maximum gap, say *maxgap*, and update it every time a new, larger gap occurs between the current prime, *p*, and the previous prime, *oldp*:

```
IF p - oldp > maxgap THEN maxgap := p - oldp;
```

You should also keep a note in the program of the *p* and *oldp* which give this maximum gap, and print them out at the end. They will be the *smallest* pair of primes which give that particular maximum gap. Check that for *max* = 1500 the maximum gap is 34, between 1327 and 1361. What is the maximum gap for *max* = 1000; for *max* = 10 000? (See again the article referred to in 1.6 above.)

1.10 A file of primes Using the program of 1.3 (or 1.4) we can produce a text file PRIMES.DAT containing the primes up to 100 000 . This will be used later in the book so we give the details.

```
PROGRAM P2_1_10;
VAR
n,q,max:   extended;
qdividesn:   boolean;
f:text;

BEGIN
assign(f, 'PRIMES.DAT');   (* If your disk is in *)
(* drive A, you may need to put 'A:PRIMES.DAT' here *)
rewrite(f);
(* Here, f is the name of the file within the program, *)
(* and PRIMES.DAT is its name on the disk *)
write ('max = ?');
readln(max);
n:=3;
writeln('prime numbers are ');
(* You might as well write primes on the screen *)
(* as well, so you can see what is happening *)
write ('2,3');
write(f,2:7);   (* These lines write 2 and 3 *)
(* to the file, leaving 7 spaces for each *)
write(f,3:7);
WHILE n <= max DO
  BEGIN
  q:=3;
```

```
REPEAT
  BEGIN
  qdividesn:=(n = q * INT(n/q));
  q:= q + 2;
  END;
UNTIL (q > SQRT(n)) OR (qdividesn);
IF NOT(qdividesn) THEN
  BEGIN
  write(',',n:0:0);
  write(f,n:7:0);    (* This writes n into the file, *)
(* leaving 7 spaces for it and no decimal places *)
  END;
n:= n+2;
END;

close(f); (* This MUST be included, to close the file *)
END.
```

It is not much use having the primes stored in a file if you can't get them out again. The method for doing this is exemplified by the following program, which reads the first m primes from the file PRIMES.DAT on disk and writes them to the screen.

1.11 Program to read from a file

```
PROGRAM P2_1_11;
VAR
f :   text;
i,m :   integer;
prime :   extended;
BEGIN
assign(f, 'PRIMES.DAT');    (* If this is in disk *)
(* drive A, then you may need to put A:PRIMES.DAT *)
reset(f);
writeln('How many primes?   ');
read(m);
FOR i:= 1 TO m DO
  BEGIN
  read(f,prime);
  write(prime:0:0,',');
  END;
close(f);
END.
```

1.12 Computing exercise   For which values of x is the prime number function $\pi(x)(=$ number of primes $\leq x)$ equal to 500? Equal to 5000?

We can now write a program which takes any number x and, if all prime factors of x are $< 100\,000$, factorizes x into its prime factors. Of course, a given x may have a prime factor p to a power > 1, so having discovered that $p \mid x$ it is still necessary to carry on dividing by p until the quotient is *not* divisible by p. The program here assumes that the primes used for trial division are available in a text file called PRIMES.DAT as in 1.10. It is also possible to replace trial division by primes with trial division by 2 and then all odd numbers up to the square root, in which case access to a file of primes is unnecessary, though the program is considerably slower.

1.13 Program for factorization by trial division

```
PROGRAM P2_1_13;
VAR
f:  text;    (* We shall assume that *)
(* the file of primes contains primes up to 100 000 *)
n, prime:  extended;
BEGIN
writeln('What is the number to be factorized?');
read(n);
assign(f,'PRIMES.DAT');
reset(f);
WHILE n > 1 DO
  BEGIN
  read(f,prime);
  WHILE (n/prime = INT(n/prime)) DO
    BEGIN
    write(prime:0:0,',');
    n:= n/prime;
    END;
  END;
close(f);
END.
```

This program writes to the screen the various prime factors of n, separated by commas. If a prime factor occurs several times in n then it is written that number of times. The program requires, of course, that *all* the prime factors of n are contained in the file of primes. If not, the program will go off the end of the file and start dividing by zero.

Perhaps you can amend the program so that an appropriate message (e.g. 'No prime factor under 100 000') is printed out if the end of the file is reached. The way to prevent the WHILE loop from running off the end of the file is to replace WHILE n > 1 DO with

WHILE ((n > 1) AND (NOT(EOF(f)))) DO where EOF(f) stands for 'end of the file called f'. Note that using the file of primes and trial division it is possible to show that a given number $< 10^{10}$ is prime, by means of trial division up to the square root: if no prime factor $< 10^5$ is found, then the number is prime. In later chapters we shall introduce a number of other techniques for proving primality, and it is good to limit proofs by trial division to some upper limit such as 10^{10}. This gives the flavour of proving big numbers are prime: really big numbers are beyond the reach of trial division proofs of primality even on very powerful computers, since this method is very lengthy. Thus special techniques, such as the ones we introduce, have to be used.

1.14 Computing exercises

(a) Verify that each of the numbers $n = 29\,341$, $314\,821$, $172\,081$ and $564\,651\,361$ has the following property: n is a product of distinct primes p, and $p-1$ is always a factor of $n-1$. (Such numbers are called *Carmichael numbers* and we shall meet them in Chapter 4, 4.8 below.)

(b) Factorize the 'repunit' r_7 whose decimal expansion consists of 7 ones. With primes up to 100 000 in the file of primes, the program fails to factorize r_{11} (11 ones) since one factor is $> 10^5$. (Compare Chapter 1, 1.5(g).)

(c) Factorize the numbers $10^n + 1$ for $n = 9, 11, 15$. (For $n = 11$ you should get $11^2 \cdot 23 \cdot 4093 \cdot 8779$.) The program will fail, however, with $n = 7$, which has a prime factor $> 10^5$.

(d) Factorize $10^{10} + 18$ and note how slow the program is.

(e) Try $n = 500\,000\,002$: this has a prime factor $> 10^5$ so the program will not guarantee that the full factorization has been obtained.

It is not difficult to modify the program 1.13 so that it can cope with numbers *one* of whose prime factors is $> 10^5$ but $< 10^{10}$. The idea is to trial-divide only by primes up to the square root of the current n, and, if no divisor is found, to declare the current n to be prime and to halt the division.

1.15 A modified trial-division program A modification of P2_1_13 along the lines suggested above is as follows.

```
PROGRAM P2_1_15;    (* Improved trial division *)
VAR
```

```
f:text; n,prime:extended;
BEGIN
writeln('Type n');
read(n);
assign(f,'PRIMES.DAT');
reset(f);
prime:=1;
WHILE ((n>1) AND (NOT(EOF(f))) AND (prime*prime <=n)) DO
  BEGIN
  read(f,prime);
  WHILE ((n/prime = INT(n/prime))
  AND (prime*prime <=n)) DO
    BEGIN
    write(prime:0:0,',');
    n:=n/prime;
    END;
  END;
IF n>1 THEN write(n:0:0);
close(f);
END.
```

If the number written out at the end is $< 10^{10}$ then we can be sure that it is prime, since it has been trial-divided by all the primes in the file, so certainly by all primes up to its square root.

1.16 Computing exercises

(a) Write r_n for the repunit whose decimal expansion is n ones. Using extended precision, r_n for $n = 7, 9, 11, 13, 15$ can all be factored completely using 1.15. Check these; for example $r_{15} = 3 \cdot 31 \cdot 37 \cdot 41 \cdot 271 \cdot 2\,906\,161$: there is only one prime factor $> 10^5$ and that is proved prime by the program since it is trial-divided by all primes up to its square root. On the other hand r_{17} has two large prime factors so the program will fail to factorize it (even assuming extended precision is valid for this number).

(b) Factorize the number $10^n + 1$ completely for $n = 12$ and $n = 13$. (For $n = 12$ you should get $73 \cdot 137 \cdot 99\,990\,001$.)

(c) Factorize completely $2^{32} + 1 = 4\,294\,967\,297$ (the fifth Fermat number).

(d) Modify the program P 2_1_15 so that it prints out nothing if n is composite and prints out n when n is prime. (So you will need to store the original n as say $n0$ and write n if $n = n0$, removing the write(prime:0:0) statement). Find the first 100 primes which have the form $k^2 + 1$ for an integer k. (The smallest is 2 ($k = 1$) and the largest is 739 601 ($k = 860$).)

(e) Try $n = 10^{10} + 18$ and note how much faster it is than the previous program P2_1_13. Try also 500 000 002 again and note that the full decomposition into primes *is* obtained since the final prime factor, 148 721, is $< 10^{10}$.

(f) Try $n = 10 113 949 037$. What is the *most* that you can deduce from the output of the program?

1.17 Special factorization related to lcm and gcd

Let r, s be given. In Chapter 1, 2.6(b), it is shown how to find u, v, x, y with $r = ux$, $s = vy$, $(u, v) = (x, y) = 1$ and $uv = \mathrm{lcm}(r, s)$, $xy = \gcd(r, s)$. We can use the file of primes to find x and y from r and s. This will be useful later, in Chapter 6, 2.11 and Chapter 8, 2.10. For example, the program could be organized as follows:

```
PROGRAM P2_1_17;
VAR
f:   text;
r,s,x,y,p,xtemp,ytemp,ri,si:   extended;
BEGIN
assign(f, 'PRIMES.DAT');
reset(f);
writeln('input r and s');
readln(r,s);
x:=1; y:=1;
WHILE ((r>1) OR (s>1)) DO
  BEGIN
  read(f,p);
  xtemp:=1; ytemp:=1; ri:=0; si:=0
  (* Here, ri and si correspond to the power of *)
  (* the prime pi in the prime-power expansions *)
  (* of r and s *)
  WHILE ((r/p) = INT(r/p)) DO
    BEGIN
    ri:=ri+1;
    r:=r/p;
    xtemp:=xtemp*p;
    END;
  WHILE ((s/p) = INT(s/p)) DO
    BEGIN
    si:=si+1;
```

```
    s:=s/p;
    ytemp:=ytemp*p;
    END;
  IF (ri>=si) THEN
    y:=y*ytemp (* Note no semicolon here !*)
  ELSE
    x:=x*xtemp;
    writeln('p=',p:0:0,' x=',x:0:0,' y=',y:0:0);
    END;
  END.
```

How would you amend this program to find u and v instead of x and y? A harder exercise is to amend it so that it finds *all* the solutions x and y (compare Chapter 1, 2.6(b)).

1.18 Project: the number of prime factors of n

For any $n \geq 1$, the function $\Omega(n)$ is defined to be the total number of prime factors of n ($\Omega(1)$ is taken to be 0). Thus $200 = 2^3 \cdot 5^2$, so $\Omega(200) = 3 + 2 = 5$. (There is also a function ω which counts the *distinct* prime factors of n : $\omega(200) = 2$.) Adapt the program 1.11 to find $\Omega(n)$. Let $f(n) = \sum_{k=1}^n \frac{(-1)^{\Omega(k)}}{k^2}$. Estimate the value of $f(n)$ as $n \to \infty$. (You can read about Ω in, for example, [Schroeder (1986), §11.3].)

1.19 Project: $4n+1$ vs $4n-1$ All primes besides 2 have one of these two forms, and the relative distributions of the two kinds make for interesting investigations. Check that $11\,593$ is the first of a sequence of nine consecutive primes of the form $4n+1$. Check that the smallest N for which (number of primes $\leq N$ of the form $4n+1$) >(number of primes $\leq N$ of the form $4n-1$) is $N = 26\,861$. (Thus primes of the form $4n-1$ tend to dominate the lists of 'small' primes.) Note that there are infinitely many primes of both sorts: see Chapter 1, 1.7, and Chapter 11, 1.5(b). You can find more information in [Riesel (1985), p. 79ff].

1.20 Project: primitive prime factors Let a be fixed and ≥ 2. Consider the sequence of numbers $a^n - 1$ for $n = 1, 2, 3, \ldots$. A prime p is called a *primitive prime factor* of $a^n - 1$ if $p \mid (a^n - 1)$ but, for all m, $1 \leq m < n$, $p \nmid (a^m - 1)$. For example, with $a = 2$, $2^4 - 1 = 15 = 3 \cdot 5$, and 3 is not a primitive prime factor since $3 \mid (2^2 - 1)$, but 5 is a primitive prime factor since 5 does not divide $1, 3$, or 7. As another example, $2^6 - 1 = 63 = 3^2 \cdot 7$, and neither 3 nor 7 is primitive since $3 \mid (2^2 - 1)$ and $7 \mid (2^3 - 1)$.

It is an astounding fact (proved by K. Zsigmondy in 1892) that the latter example is the *only* one where no prime factor is primitive. In fact it was proved by A. Schinzel in 1962 that most numbers in a sequence as above have at least two primitive prime factors. See [Ribenboim (1988), p. 32].

Write a program to determine primitive prime factors of sequences $a^n - 1$. Find, for $a = 2, 3$ and 4, all those numbers in the sequence which have just *one* primitive prime factor, taking the calculations up to some reasonable limit. (Let us note in passing that PASCAL has no exponentiation operation: to work out a^n exactly you need to multiply a by itself repeatedly. That is no hardship when you want in any case to consider a^m for values of $m < n$.)

1.21 An alternative file type

It is possible to store the primes in a file of extended precision numbers (or indeed reals) and this needs only very small modifications to the programs given above. Here is a suitable program to produce the primes $\leq n$.

```
PROGRAM P2_1_21;
VAR
x,y,n:  extended;
ydividesx:  boolean;
f:  file of extended;
BEGIN
   assign(f, 'smallprimes'); (* Just to change the name *)
   rewrite(f);
   writeln('n = ?');
   readln(n);
   writeln('prime numbers are ');
   x:= 2;
   write(f,x);
   (* This is the way to write 2 into the file *)
   write(2,',');
   (* So 2 is written to the screen too *)
   x:= 3;
   write(f,x);
   write(3);
   WHILE x <= n DO
     BEGIN
       y:= 3;
       REPEAT
         BEGIN
```

```
            ydividesx:= (x = y* INT(x/y));
            y:= y+2;
          END;
        UNTIL (y > SQRT(x)) OR (ydividesx);
        IF NOT (ydividesx) THEN
          BEGIN
            write(f,x);
            write(',',x:0:0);
          END;
         x:=x+2;
      END;
    close(f);
  END.
```

There is one small advantage to this kind of file, and that is that it is possible to pull out any given element of the file using the SEEK procedure. Here is a suitable program for this.

```
PROGRAM P2_1_21A;
VAR
k:longint;
n:extended;   (* Or integer and real will do *)
f:   file of extended;
BEGIN
  assign(f, 'smallprimes');
  reset(f);
  writeln('type the value of k');
  (* The k-th element will be pulled out *)
  readln(k);
  seek(f,k - 1);
  (* Why k - 1 ?  Because files are numbered from 0 !  *)
  read(f,n);    (* So the k-th element is called n *)
  write(n:0:0);
  close(f);
END.
```

1.22 Project It can be shown that there is an integer $N \le 100$ such that, for all $n \ge N$, *at least one* of the numbers $n, n+1, n+2, \ldots, n+9$ has three or more distinct prime factors. Write a program to make a good guess at the number N. [See *Amer. Math. Monthly* **96**(1989), 942-4, for the proof.]

2. Primes by multiplication (elimination of composites)

We turn now to the second method of listing primes. This one is particularly suited to listing primes p in some interval $r < p \leq r+2n-1$ where r is even and n is any positive integer. As before, we shall consider only *odd* numbers as candidates for p, and we ask the question: when can a composite number ij (i and j odd and > 1) be of the form $r + 2k - 1$ for some k with $1 \leq k \leq n$? When these composites are eliminated, the numbers which remain must be prime!

Write $c = r + 2n - 1$. With $r + 1 \leq ij \leq c$ we get $\frac{r+1}{i} \leq j \leq \frac{c}{i}$. Now the smallest odd integer $\geq \frac{r+1}{i}$ is (see Chapter 1, 4.3(j))$s = 2\left[\frac{r+i}{2i}\right] + 1$. So this is the smallest value of j, unless $s = 1$ in which case we replace s by 3. The range of values of i that we need to consider is $3 \leq i \leq m$, where $m = [\sqrt{c}]$. For any composite number $\leq c$ will have a factor $\leq \sqrt{c}$ (compare 1.1 above).

For a given i, we start with $j = s$ and increment j in steps of 2 to $[\frac{c}{i}]$; for each such j we work out ij. When $ij = r + 2k - 1$, we have $k = \frac{1}{2}(ij - r + 1)$, so this value of k is eliminated. The values of k in the interval $1 \leq k \leq n$ which remain are those which give the primes $r + 2k - 1$. To keep track of those k which are eliminated we start by setting up an array of integers, say $a[1..n]$, and set them initially all to 0 (PASCAL will not do this for us if we say nothing about the initial values). Then we change $a[k]$ to 1 if it is eliminated as above, and print out the numbers $r + 2k - 1$ for those k having $a[k] = 0$ at the end.

It is advisable to work through a simple example by hand before trying the program below. Try $r = 30$, $n = 15$, so $c = 59$, $m = 7$ and we are looking for products ij with $31 \leq ij \leq 59$ and $3 \leq i \leq 7$. For $i = 3$, the values of j are $11 \leq j \leq 19$ (odd j only, of course), giving $ij = 33, 39, 45, 51, 57$, and eliminating $k = 2, 5, 8, 11, 14$. For $i = 5$ we have only $j = 7, 9, 11$ and for $i = 7$ we have only $j = 5$ and 7. Naturally, some numbers are eliminated twice; for example 45 ($k = 8$) goes out with $i = 3$, $j = 15$ and with $i = 5$, $j = 9$.

2.1 Program for primes by multiplication

```
PROGRAM P2_2_1;
VAR
r,c,m,s,i,j:  extended;
n,k:  integer;
key:array[1..10000] of integer;
(* So n should not be larger than 10000 *)
BEGIN
```

```
writeln('What is the value of r (even)?   '); readln(r);
writeln('What is the value of n (<= 10000)'); readln(n);
c:= r + 2*n + 1;
m:= INT(SQRT(c));
FOR k:= 1 TO n DO
  key[k]:= 0;
i:= 3;
WHILE i <= m DO
  BEGIN
    s:= 2*INT((r+i)/(2*i)) + 1;
    IF s = 1 THEN s:= 3;
    j:= s;
    WHILE j <= INT(c/i) DO
      BEGIN
        k:=TRUNC((i*j - r + 1)/2);
        (* This makes k an integer *)
        key[k]:= 1;
        j:= j + 2;
      END;
    i:= i + 2;
  END;
FOR k:= 1 TO n DO
  BEGIN
    IF key[k] = 0 THEN
    write((r + 2*k - 1):0:0,',');
  END;
END.
```

2.2 Computing exercise   Verify that there are 11 primes between 10^4 and $10^4 + 100$. Find the primes between 10^n and $10^n + 100$ for $n = 5, 6, 7, 8$. Do the same for primes between 10^n and $10^n + 1000$ (for $n = 8$ there are 54 such primes).

3. Approximations to $\pi(x)$

The number $\pi(x)$, which is defined for all real x, is the number of primes $\leq x$. The history of number theory is punctuated with attempts to find a good approximation to the function π. There are several approximations now known, and the relationships between them, and the exactness with which they approximate π, make fascinating reading. We must refer the reader to [Riesel (1987), Ch. 2] or [Ribenboim (1988), Ch. 4] for the details, which are too technical for this book. What we can do here is to state the various formulas and invite you to compare them with the true values. In the next section we shall in fact give a (rather slow) method by which $\pi(x)$ can be calculated for any given x, without actually finding all the primes $\leq x$. But the method of Section 2 above enables us to find the number of primes in a given interval and that is a more rapid way of comparing fact with formula, for we can look at relatively large primes around the value x without having to wait all day for the total number less than x to be calculated.

3.1 $\pi(x)$ **is approximately** $x/\log x$ The precise statement is that $\pi(x)\log x/x \to 1$ as $x \to \infty$. This was conjectured by Gauss (and in a slightly different form by Legendre) and proved independently by J. Hadamard and C. J. de la Vallée-Poussin in 1896. (Note that log is the natural logarithm, to the base e.) One way of thinking of the formula is that about 1 in $\log x$ of the numbers up to x are prime, when x is large. So, for example, around $x = 10^6$, we expect about 1 in 14 numbers to be prime, and around $x = 10^{50}$, we expect about 1 in 115 numbers to be prime. The primes thin out according to a logarithmic law, but of course this is only an *average* thinning out: it is also believed that there are infinitely many pairs of primes differing by two!

Using this formula, the number of primes between x and y is approximately $\frac{y}{\log y} - \frac{x}{\log x}$. Putting $y = 10^4 + 100$ and $x = 10^4$ this comes to 9.67; as in 2.2 above there are actually 11 primes in this interval. This approximation is not at all bad bearing in mind that there are 1229 primes below x and 1240 primes below y.

3.2 Computing exercise   Compile tables of the estimates for the number of primes in intervals of the form 10^n to $10^n + 10^k$, where $k = 2$, 3 or 4, and compare them with the true values as calculated by the method of Section 2.

We shall mention two other approximations to $\pi(x)$ here, the *logarithmic integral* and *Riemann's function*.

3.3 Computing exercise: the logarithmic integral   This is defined as

$$Li(x) = \int_2^x \frac{dt}{\log t},$$

where as usual log means the natural logarithm. Write a program to calculate this integral by Simpson's rule. The second version of the prime number theorem states that $\pi(x)/Li(x) \to 1$ as $x \to \infty$. Apply this to the same examples as in 3.2 and compare the results obtained. Once a means is found, in Section 4 below, for calculating $\pi(x)$ exactly for some fairly large x, further comparisons can be done (see 'Legendre's sieve' and 4.10 below). [In case you have forgotten Simpson's rule, it states the following approximation to $I = \int_a^b f(x)\mathrm{d}x$. Divide the interval into $2n$ equal pieces by $x_0 = a$, $x_1 = a + h$, $x_2 = a + 2h, \ldots, x_{2n} = a + 2nh = b$, where $h = (b-a)/2n$. Then $I \approx \frac{1}{3}h(y_0 + y_{2n} + 2(y_2 + y_4 + \ldots + y_{2n-2}) + 4(y_1 + y_3 + \ldots + y_{2n-1}))$, where $y_i = f(x_i)$.]

3.4 Project: Riemann's function   A slightly more ambitious exercise is to use the function

$$R(x) = Li(x) - \frac{1}{2}Li(x^{1/2}) - \frac{1}{3}Li(x^{1/3}) - \frac{1}{5}Li(x^{1/5}) + \ldots .$$

(You will no doubt be intrigued by the question of which term comes next! See for example [Riesel (1987), p. 54] for further information on $R(x)$.) It is also true that $\pi(x)/R(x) \to 1$ as $x \to \infty$, and, for values of x within the reach of extended precision, $R(x)$ is a considerably better approximation to $\pi(x)$ than either of the previous approximations. Experiment with the values obtained before, using just the first few terms of the expansion for $R(x)$ given above. Again further corroboration can be sought at a later stage: see 4.10 below.

3.5 Times for primality by trial division   It is a sobering exercise to use the prime number theorem 3.1 for estimating the time needed to prove primality by trial division. To prove a number n is prime it is necessary to trial-divide it by all primes $\leq \sqrt{n}$, and there are about $\sqrt{n}/\log\sqrt{n}$ of these. Taking $n = 10^k$ this means trial-dividing by about $10^{k/2}/((k/2)\log 10)$ numbers. Let us assume, very optimistically, that 10^6 of these trial divisions can be done every *second*; then the time to prove primality will be about

$$\frac{10^{k/2}}{(k/2)\log 10 \times 10^6 \times 60 \times 60 \times 24 \times 365} = \left(\frac{10^{k/2}}{k}\right) \times 2.75 \times 10^{-14} \text{ years.}$$

Even for the very modest $k = 30$ (30-digit primes) this comes to around 11 months, while for $k = 50$ it comes to a staggering 5.5×10^9 years. For $k = 100$ it is more like 2.75×10^{34} years. So primality proofs by trial division are not very feasible for large numbers. In this book we have restricted such proofs to numbers $< 10^{10}$ in order to give some flavour of proving numbers are prime by other methods. What the calculations here show is that you do not have to go very much higher than that before proofs by trial division are impractical even on the largest computers.

4. The sieve of Eratosthenes

The final method we shall study for listing primes is called a sieve, and is attributed to Eratosthenes around 200 B.C. We start with the integers ≥ 2 in a list:

2	3	4	5	6	7	8	9	10	11	12	13	14	15	16	17	18
		×		×		×		×		×		×		×		×
				×			×			×			×			×
								×					×			

19	20	21	22	23	24	25	26	27	28	...
	×		×		×		×		×	
		×			×			×		
	×					×				

The first step is to 'sieve by 2', i.e., remove all multiples of 2 besides 2 itself. The first row of crosses does this. The smallest unsieved number > 2 is now 3, and we sieve by 3, i.e., remove all multiples of 3 besides 3 itself. The second row of crosses does this. The smallest unsieved number > 3 is now 5 and we repeat by sieving with 5. In this way, all composite numbers are sieved out (some several times) and what remain are primes. Notice that in the example above, containing numbers up to 28, sieving by 2, 3 and 5 leaves only primes. In fact, the first composite that would be left is $49 = 7^2$, since any smaller number has a prime factor $\leq [\sqrt{48}] < 7$, and therefore ≤ 5. Sieving by $2, 3, 5$ and 7 will remove all composites $< 11^2 = 121$.

The sequence of smallest numbers which remain after the succession of sievings are consecutive primes starting with 2. (This is evident from the example above, but if you want to see it formally then an induction argument is required. It is certainly true after one sieving, since 2 and 3 are the numbers in question. Assume that the first k sievings are by the first k primes $2, 3, \ldots, p_k$ and that the smallest unsieved number $> p_k$ is p_{k+1}, the $(k+1)$st prime. Now sieve by p_{k+1}, i.e., remove all multiples of p_{k+1} besides p_{k+1} itself. The next prime, p_{k+2}, will certainly not be sieved out, but all remaining composites between p_{k+1} and p_{k+2} will be, since they are divisible by *some* prime $< p_{k+2}$. This completes the induction.)

We can expedite the sieve by a few simple tricks. When sieving by a prime $p > 2$, the numbers $2p, 3p, \ldots, (p-1)p$ will have gone at an earlier stage (being divisible by some prime $\leq \sqrt{(p-1)p}$ and therefore $< p$). So the first number which p sieves and which was not already sieved is

p^2. However $p^2 + p$, $p^2 + 3p$, $p^2 + 5p$, ..., being even, have already been sieved out by 2, so need not concern us. Thus we can start the sieving by p with the elimination of p^2 and continue in steps of $2p$. (Even this is not perfectly efficient; for example $45 = 3^2 + 12 \cdot 3 = 5^2 + 4 \cdot 5$ will still get sieved twice.)

In order to be certain of finding all primes $\leq N$ we need only sieve with primes $\leq \sqrt{N}$.(E.g., for $N = 10000$ we need only sieve with primes ≤ 100.) The most straightforward way to do this is to insert these as an array of constants at the beginning of the program; alternatively they can be accessed from an already existing file of 'small' primes.

Here is a program for producing the primes ≤ 2000 by a sieve. We shall sieve only by odd primes, so assume that even numbers have been sieved out already. But to simplify things we set up an array key[1..2000] of booleans, $key[i]$ set initially to TRUE and changed to FALSE when i is sieved out. At the end we print out all the odd i for which $key[i]$ remains TRUE. (Of course, you can equally well have an array of integers, giving $key[i]$ the value 1 at the beginning and changing it to 0 if i is sieved out.)

4.1 Program for the sieve of Eratosthenes

```
PROGRAM P2_4_1;
CONST
f:  array[0..12] of integer = (3, 5, 7, 11, 13, 17,
        19, 23, 29, 31, 37, 41, 43);
VAR
i,j,prime:  integer;
key:  array[1..2000] of boolean;
BEGIN
  FOR i:= 1 TO 2000 DO
    key[i]:= TRUE;
  j:= 0;
  REPEAT
    BEGIN
      prime:=f[j];
      i:= prime*prime;
      WHILE i <= 2000 DO
        BEGIN
          key[i]:=FALSE;
          i:=i+2*prime;
        END;
      j:= j + 1;
    END;
```

```
    UNTIL (prime = 43);
    writeln('primes less than 2000 are:- 2');
    i:= 3;
    WHILE i <= 2000 DO
      BEGIN
        IF key[i] THEN
          write(',',i);
        i:= i + 2;
      END;
  END.
```

4.2 Computing exercises

(a) Adapt the program 4.1 to produce primes ≤ 5000.

(b) Make the program 4.1 slightly more efficient by ignoring even numbers completely, that is by using an array key[1..999] corresponding to the odd numbers $3, 5, 7, \ldots, 1999$. At the end, $2i + 1$ is prime if and only if key[i] $=$ TRUE.

It is possible to be considerably more subtle in writing a sieving program, in effect creating small primes and continually using them to create bigger ones by sieving. Here is a sample program; see what you can make of it! It sieves the odd numbers from 3 to $2n + 1$ and $a[k]$ is false if and only if $2k + 1$ is found to be composite.

4.3 Alternative program for the sieve

```
  PROGRAM P2_4_3;
  CONST
  n=1000;
  VAR
  p,q,i,j,k:  longint;
  (* Better than integer because q gets big *)
  a:  array[1..n] of boolean;
  BEGIN
    FOR j:=1 TO n DO
      a[j]:= true;
    p:= 3; q:= 4; j:=1;
    WHILE j<=n DO
      BEGIN
        IF a[j]= true THEN
          BEGIN
            write(p, ',');
            k:=q;
```

```
            WHILE k<=n DO
              BEGIN
                a[k]:=false;
                k:=k+p;
              END;
          END;
        j:= j+1; p:= p+2; q:=q+2*p- 2;
        (* Think of 2q+1=p² *)
      END;
    END.
```

It is of course possible to modify the program so that it only sieves with primes $\leq \sqrt{2n+1}$. Try it!

We shall now show how to use the sieve to *count* the primes $\leq x$ by assuming a knowledge only of the primes $\leq \sqrt{x}$. Items 4.4 to 4.10 below can be regarded as optional reading. The sieve tells us that this must in principle be possible, since the latter primes determine the primes $\leq x$. The method is due to A.M. Legendre (1752–1833), and is sometimes known as the *Legendre sieve*. Many sophisticated variants have been found, by E.D.F. Meissel (culminating in a wrong value for $\pi(10^9)$ published in 1885), D.H. Lehmer (1959) and later authors (see [Riesel (1987), pp.13ff]). Using these, some astounding results have been achieved, such as

$$\pi(4 \cdot 10^{16}) = 1\,075\,292\,778\,753\,150,$$

in 1985. Remember this is an *exact* value. See [Ribenboim (1988), p. 178]. The general idea is to use the primes $\leq \sqrt[3]{x}$ or even $\sqrt[4]{x}$ to count the primes up to x. This involves some subtle ideas and we shall not pursue them here.

4.4* Example of Legendre's sieve We begin with a simple example: let us count the primes ≤ 28 by using the primes $\leq \sqrt{28}$, i.e. the primes 2, 3 and 5. Let C_i, $i = 1, 2, 3$ (one for each prime) be, respectively, the multiples of 2, 3 and 5 in the set $X = \{2, 3, \ldots, 28\}$. The sets C_i and their overlaps are shown in Fig. 2.1. The primes ≤ 28 are, of course, precisely the numbers outside the union $C_1 \cup C_2 \cup C_3$, together with the original primes 2, 3 and 5. Note that $C_1 \cap C_2 \cap C_3$ is empty since $28 < 2 \cdot 3 \cdot 5$.

We use the following notation: #A denotes the number of elements in a (finite) set A, $N(i) = \#C_i$, $N(i,j) = \#(C_i \cap C_j)$ for $i \neq j$, and so on.

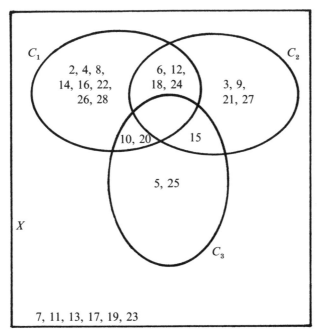

Fig. 2.1

Thus

$$N_1 = \text{Number of elements of } X \text{ divisible by } 2 = [\tfrac{28}{2}] = 14,$$

$$N(1,2) = \text{Number of elements of } X \text{ divisible by } 2 \text{ and } 3$$

$$= \text{Number of elements of } X \text{ divisible by } 2 \cdot 3 = [\tfrac{28}{2.3}] = 4.$$

and so on.

Finally, we define $S_0 = \#X = 27$,

$$S_1 = N(1) + N(2) + N(3) = [\tfrac{28}{2}] + [\tfrac{28}{3}] + [\tfrac{28}{5}] = 14 + 9 + 5 = 28,$$

$$S_2 = N(1,2) + N(2,3) + N(3,1) = [\tfrac{28}{6}] + [\tfrac{28}{15}] + [\tfrac{28}{10}] = 4 + 1 + 2 = 7.$$

In this example there is no S_3 since $N(1,2,3) = 0$.

The *principle of inclusion–exclusion* enables us to count the number of elements of X not belonging to any of the C_i: it is the alternating sum $S_0 - S_1 + S_2 = 27 - 28 + 7 = 6 : \pi(28) - \pi(\sqrt{28}) = 6$.

More generally let $X = \{2, 3, \ldots, [x]\}$ for any x, let $p_1, p_2, p_3, \ldots$ be the primes in ascending order $\leq [\sqrt{x}]$ and let $C_i = \{a \in X : p_i \mid a\}$. For any indices $i_1 < i_2 < \ldots < i_k$ let $N(i_1, i_2, \ldots, i_k) = \#(C_{i_1} \cap C_{i_2} \cap \ldots \cap C_{i_k})$. Since the p_i are distinct primes, this is precisely the number of elements of X divisible by the *product* $p_{i_1} p_{i_2} \ldots p_{i_k}$

(compare 1.2 or 2.3(a) in Chapter 1). Consequently,

$$N(i_1, i_2, \ldots, i_k) = \left[\frac{x}{p_{i_1} p_{i_2} \cdots p_{i_k}} \right].$$

As in the example above, we define $S_0 = \#X = [x] - 1$ and, for $k \geq 1$,

$$S_k = \sum N(i_1, i_2, \ldots, i_k)$$

where the summation is over all k-tuples $i_1 < i_2 < \ldots < i_k$ for which the corresponding primes are ≥ 2 and $\leq [\sqrt{x}]$.

The number N above will be zero whenever the product of primes in the denominator exceeds x, and this will happen as soon as the product of the *first* k primes $p_1 p_2 \ldots p_k$ exceeds x, for the product in the denominator is $\geq$ this product. So the largest value of k which occurs is say K where

$$p_1 p_2 \ldots p_K \leq [x] < p_1 p_2 \ldots p_{K+1}.$$

The principle of inclusion–exclusion then says the following.

4.5* Principle of inclusion–exclusion   *The number of elements of X which do not lie in any of the C_i is*

$$S_0 - S_1 + S_2 - \ldots + (-1)^K S_K.$$

Proof   Consider an integer v which belongs to precisely n (≥ 1) of the sets C_i. How many times is v counted in the various sums S? It is counted once in S_0, of course. In S_1 v is clearly counted n times, once for each of the n sets to which it belongs. In S_2 every *pair* of sets is used, and so every time two sets are chosen from among the n, v is counted once: a total of $\binom{n}{2}$ times. Similarly in S_3, v is counted $\binom{n}{3}$ times, and so on. Thus in the alternating sum of S_i, v is counted precisely

$$1 - \binom{n}{1} + \binom{n}{2} - \ldots + (-1)^n \binom{n}{n}$$

times. But if $n \geq 1$ this is $(1-1)^n = 0$. On the other hand if v belongs to *none* of the C_i then it is counted once only, in S_0.   □

In our case the elements of X which do not lie in any of the C_i are those numbers $\leq [x]$ which are not divisible by any prime $\leq [\sqrt{x}]$, that is they are the primes $> [\sqrt{x}]$ and $\leq [x]$. Hence:

4.6* Theorem $\pi(x) - \pi(\sqrt{x}) = S_0 - S_1 + S_2 - \ldots + (-1)^K S_K.$

4.7* Exercise   Apply the theorem to $x = 100$, so that the primes $\leq [\sqrt{x}] = 10$ are 2,3,5 and 7. For example,

$$S_2 = \left[\frac{100}{2 \cdot 3} \right] + \left[\frac{100}{2 \cdot 5} \right] + \ldots + \left[\frac{100}{5 \cdot 7} \right].$$

Since $2 \cdot 3 \cdot 5 \leq 100 < 2 \cdot 3 \cdot 5 \cdot 7, K = 3$ in 4.6. Thus find $\pi(100)$ from $\pi(10) = 4$.

You may notice that the amount of calculation needed to use 4.6 is quite large, even if it does only depend on knowing the primes up to 10. As mentioned above, there exist improvements in 4.6 which are too technical to go into here. We shall content ourselves (and we hope the reader) by implementing 4.6 in a program so that you can calculate some values of $\pi(x)$ where x has a fairly modest size. It is possible to be much more cunning and efficient in the calculation of the S_i, and you can read an account of such matters in [Riesel (1987), pp.16ff].

Consider for example the problem of finding S_4 when $x = 800$. We thus want to consider products of four primes each of which is $\leq \sqrt{800}$, that is four primes all ≤ 23. We enumerate these products as follows:

$2 \cdot 3 \cdot 5 \cdot 7$				$= 210$
$2 \cdot 3 \cdot 5 \cdot$	11			$= 330$: *increase the top prime one step*
$2 \cdot 3 \cdot 5 \cdot$		13		$= 390$
$2 \cdot 3 \cdot 5 \cdot$			17	$= 510$
$2 \cdot 3 \cdot 5 \cdot$				19 $= 570$
$2 \cdot 3 \cdot 5 \cdot$				23 $- 690$: *23 is the largest prime*

Now increase the next-to-top prime one step and drop the top prime

$2 \cdot 3 \cdot$	$7 \cdot 11$			$= 462$
$2 \cdot 3 \cdot$	$7 \cdot$	13		$= 546$
$2 \cdot 3 \cdot$	$7 \cdot$		17	$= 714$
$2 \cdot 3 \cdot$	$7 \cdot$			19 $= 798$: *next product with 23 is > 800*

Increase the next-to-top prime one step and drop the top prime

$2 \cdot 3 \cdot$	$11 \cdot 13$	$= 858$: *already too big, so increase the third-from-top prime and drop the two above it:*
$2 \cdot$	$5 \cdot 7 \cdot 11$	$= 770$: *next product with 13 too big, so increase the fourth-from-top prime (i.e. the bottom prime!):*
	$3 \cdot 5 \cdot 7 \cdot 11$	*:already too big, so that is the end.*

Having enumerated all these, the term S_4 is then the sum

$$[\tfrac{800}{2 \cdot 3 \cdot 5 \cdot 7}] + \ldots + [\tfrac{800}{2 \cdot 5 \cdot 7 \cdot 11}].$$

So the aim of the enumeration is not to waste time looking at products which are bigger than those already known to be too big.

Here is a PASCAL program which does the above enumeration, assuming that $x < 2 \cdot 3 \cdot 5 \cdot 7 \cdot 11 \cdot 13 \cdot 17 \cdot 19 = 9\,699\,690$. This means that only $S_0, \ldots, S_7$ come into the calculation and at most $\pi(\sqrt{9\,699\,690}) = 443$ primes will be needed in the sieve.

4.8* Program for Legendre's sieve

```
PROGRAM P2_4_8;
CONST
max=443;    (* See above *)
VAR
f:text;    (* This is the file of primes *)
i,p:integer;
(* i is a counting variable; p is a prime ≤ x *)
s0,s1,x,rootx,total,ans:extended;
primarr:array[1..max] of extended;
finish:boolean;

    FUNCTION sieve(no_terms:integer):extended;
    (* computes $S_{no\_terms}$ *)
    (* for the Legendre's sieve method *)
    VAR
    cond,cond4,exit:boolean; prod,s:extended; i,j:integer;
    count:array[1..10] of integer;    (* count[i]=j means *)
    (* that the j-th prime in the usual ascending *)
    (* sequence of primes 2,3,5,... occurs in the *)
    (* i-th position from the left in one of the *)
    (* products making up S *)

    PROCEDURE test(VAR prod:extended; VAR cond,
  exit:boolean);
    VAR
    i:integer; cond1,cond2,cond3:boolean;
    BEGIN
      prod:=1;
      FOR i:=1 TO no_terms DO
        BEGIN
          prod:=prod*primarr[count[i]];
          cond1:=prod > x;
          cond2:=primarr[count[no_terms]] > rootx;
          (* This means that the largest prime in *)
          (* the product is > √x *)
          cond:=cond1 OR cond2
        END;
      cond3:=count[no_terms]=count[1]+(no_terms-1);
      (* This says that the primes occurring in the *)
      (* product are consecutive primes *)
      IF cond AND cond3 THEN
```

```
      exit:=TRUE    (* i.e.  if the primes are *)
   (* consecutive and the product or the largest *)
   (* prime is too big then stop calculating S *)
END;   (* of test *)

BEGIN   (* the function 'sieve' *)
  exit:=FALSE; prod:=1; s:=0;
  FOR i:=1 TO no_terms DO
    count[i]:=i;
    (* So the first i primes are used in the product *)
  test(prod,cond,exit);
  WHILE NOT(exit) DO
    BEGIN
      WHILE cond AND NOT(exit) DO
        BEGIN
          j:=2;
          WHILE (j<= no_terms-1) AND NOT(exit) DO
            BEGIN
              cond4:=count[no_terms]=count[j]+
              (no_terms-j);
              (* This says that the primes except *)
              (* the last j- 1 form a block of *)
              (* consecutive primes *)
              IF cond AND cond4 THEN
                BEGIN
                  count[j-1]:=count[j-1]+1;
                  FOR i:=j TO no_terms DO
                    count[i]:=count[i-1]+1;
                  test(prod,cond,exit);
                  j:=no_terms
                END
          (* This finds the longest block of *)
          (* consecutive primes and moves up the first *)
          (* prime below that block *)
              ELSE
                  j:=j+1;
              END;
          END;
      WHILE NOT(cond) DO
        BEGIN
          s:=s+INT(x/prod);
```

```
            (* A new term is added to the sum for S *)
            count[no_terms]:=count[no_terms]+1;
            (* Move last prime up one *)
            test(prod,cond,exit);
            IF cond THEN
              BEGIN
                count[no_terms-1]:=count[no_terms-1]+1;
                count[no_terms]:=count[no_terms-1]+1;
                (* These move the second-to-last *)
                (* prime up one and put the last prime *)
                (* adjacent to it *)
                test(prod,cond,exit);
              END;
          END;
      END;
    sieve:=s;
  END;    (* of the function 'sieve' *)

BEGIN    (* of main program *)
  assign(f,'PRIMES.DAT');    (* This uses the file of *)
  (* primes to read in the primes up to √x *)
  reset(f);
  writeln('Program to compute the number
    of primes <= x=9,699,690');
  writeln('using the Legendre's sieve technique');
  write('x= ?  '); readln(x);
  rootx:=sqrt(x);
  FOR i:=1 TO max DO
    read(f,primarr[i]);    (* This reads the first 443 *)
      (* primes from the file into an array *)
  total:=0; s0:=INT(x)-1; writeln('s0= ',s0:8:0);
  i:=1; s1:=0;
  WHILE primarr[i] <= rootx DO
    BEGIN
      s1:=s1+INT(x/primarr[i]);
      i:=i+1;
    END;
  writeln('s1= ',s1:8:0);
  total:=s0-s1;
  finish:=FALSE;
  WHILE NOT(finish) DO
```

```
BEGIN
   i:=2;
   WHILE (i<= 7) OR (NOT finish) DO
      BEGIN
         ans:=sieve(i);
         IF ans=0 THEN
            finish:=TRUE
         ELSE
            BEGIN
               writeln('s',i:1,'= ',ans:8:0);
                  (* This prints Sᵢ *)
               IF ODD(i) THEN
                  ans:=-ans;
               total:=total+ans;
            END;
         i:=i+1;
      END;
   END;
   p:=0; i:=1;
   WHILE primarr[i]<= rootx DO
      BEGIN
         p:=p+1;
         i:=i+1;
      END;
   writeln(' number of primes up to rootx = ',p:8);
   write('the number of primes up to ',x:8:0);
   writeln(' is ',total+p:8:0);
   close(f);
END.
```

4.9* Computing exercise   Use Legendre's sieve to find $\pi(x)$ where $x = 10^5, 2 \cdot 10^5, 3 \cdot 10^5, \ldots, 10^6$. For $x = 6 \cdot 10^5$, the numbers S_i should come to: 599 999, 1 297 809, 1 263 496, 698 663, 207 842, 26 863, 961, 2; also $\pi(\sqrt{x}) = 137$, giving altogether $\pi(x) = 49\,098$.

4.10* Project: $Li(x)$ and $R(x)$ again   Find the functions $Li(x)$ and $R(x)$ (see 3.3, 3.4 above) for the values of x in 4.9. Find also $x/\log x$ for these x, and plot graphs of x against these three functions and also the true prime-number function $\pi(x)$. Which approximation is consistently better than the others?

There is a fascinating way of calculating the sum which occurs in

Legendre's sieve, by a recursive formula. Unfortunately it is even slower than the method of program 4.8, but it deserves mention since it *can* be made efficient by a lot more work. (See [Riesel (1987), p. 16].) Let $\phi(x, a)$, for real x and integral $a \geq 1$, denote the number of positive integers $n \leq x$ which are not divisible by any of the first a primes $p_1, p_2, \ldots, p_a$.(Here $p_1 = 2, p_2 = 3$, etc.) Thus when $a = \pi(\sqrt{x})$ the integers in question are 1 and the primes $> \sqrt{x}$ but $\leq x$:

$$\phi(x, a) = \pi(x) - \pi(\sqrt{x}) + 1,$$

so that $\pi(x) = \phi(x, a) + a - 1$, when $a = \pi(\sqrt{x})$.

Now for $a > 1$, $\phi(x, a) = \phi(x, a - 1) - \#\{n \leq x : p_1 \nmid n, \ldots, p_{a-1} \nmid n, p_a \mid n\}$. This latter set is in one-to-one correspondence with the set of integers $\leq x/p_a$ and not divisible by any of $p_1, \ldots, p_{a-1}$ (via the correspondence $n \to n/p_a$). Hence we have the recursive formula:

4.11 $\phi(x, a) = \phi(x, a - 1) - \phi(x/p_a, a - 1)$.

Here is a program to use this for calculating $\phi(x, a)$. The value of $\phi(x, 1)$ is simply the number of odd numbers $\leq x$. You should check that this equals $[(x + 1)/2]$.

4.12 Program for $\phi(x, a)$

```
PROGRAM P2_4_12;
VAR
b,k:  integer;
y:  real;
f :  text;
p:  array[1..100] of integer;   (* This will be an *)
(* array of primes, drawn from the file PRIMES.DAT. *)
(* Note that this array is available to the function *)
(* phi without further declaration *)

  FUNCTION phi(x:real; a:integer):integer;
  BEGIN
    IF a > 1 THEN
      phi:= phi(x, a - 1) - phi(x/p[a], a - 1)
    ELSE
      phi:= trunc((x + 1)/2);
  END;

BEGIN   (* main program *)
  assign(f, 'PRIMES.DAT');
  reset(f);
  FOR k:= 1 TO 100 DO
```

```
   read(f, p[k]);
   close(f);

   writeln('Type the values of x and a separated by a
   space');
   readln(y,b);
   writeln(phi(y,b));
END.
```

4.13 Computing exercise   Use the above program to find $\pi(1000)$, $\pi(2000)$, $\pi(3000)$ and $\pi(4000)$. You may notice that it is getting quite slow by this time. Try doing a much smaller example by hand to see why it is so slow. Can you improve the speed of the program?

4.14 Project: lucky numbers   There are other sieves which resemble the sieve of Eratosthenes and which produce other sequences of numbers with a distribution rather like the primes. In fact it could be said that the distribution of primes owes less to their 'primeness' than to the fact that they can be generated by a sieve. Here is one such sieve, which is described in, for example, [Gardiner *et al.* (1956), Hawkins and Briggs (1957/8)].

Start as before with the positive integers. Begin by crossing out every second integer in the list, starting the count with 1 (so you cross out $2, 4, 6, \ldots$). Other than 1 the smallest integer left is 3, so we continue by crossing out every third integer left, starting the count with 1 (cross out 5, 11, 17, ...). Now 7 is the smallest integer left after 3, so cross out every seventh integer left, starting the count with 1. This is continued indefinitely.

The integers which remain are the *lucky numbers*. Write a program to find all the lucky numbers $< 10\,000$. Verify the following figures for the number $L(x)$ of lucky numbers $\leq x$:

x	10^3	$2 \cdot 10^3$	$3 \cdot 10^3$	$4 \cdot 10^3$	$5 \cdot 10^3$	$6 \cdot 10^3$	$7 \cdot 10^3$	$8 \cdot 10^3$	$9 \cdot 10^3$	10^4
$L(x)$	153	276	390	503	610	716	816	920	1019	1118

How does this fit with a 'lucky number theorem' of the form $L(x) \approx x/\log x$? You will find other data on the lucky numbers in the first paper cited above.

Here are a few hints on writing the program. It is a mistake to use a single array to contain, say, the odd numbers up to $10\,000$ and to successively sieve through this array eliminating more numbers on each pass. This involves sieving through nearly the whole array each time (note that with primes, we could at least start with p^2 where p was the current prime being used for sieving). It is better to use two arrays, the first containing the current numbers which have survived sieving, and with a zero at the end of the numbers to act as a 'sentinel' marking the end. The second array will contain only the numbers which survive the sieving which takes place at the current pass through the first array. Thus numbers are written from the first array to the second only when they survive sieving. For example, having sieved with 3 the numbers in the first array will be

$$1, 3, 7, 9, 13, 15, 19, 21, 25, 27, 31, 33, 37, 39, 43, 45, \ldots \quad .$$

The numbers written into the second array are then (sieving by 7)

$$1, 3, 7, 9, 13, 15, \; 21, 25, 27, 31, 33, 37, 43, 45, \ldots \quad .$$

The roles of the two arrays are then changed over and the next pass, sieving by 9, is performed.

3

Congruences

1. Congruences: basic properties

The idea behind the congruence notation is very simple, but it is so useful that we shall use it in practically every page of this book from now on. The notation was introduced by C.F. Gauss (1777–1855) in his seminal work *Disquisitiones Arithmeticae* which was published in 1801. It is still a wonderful and lucid account of the subject and is available in English translation [Gauss (1986)].

1.1 Definition   *Let a, b and m be integers, with $m \neq 0$. We define $a \equiv b \pmod{m}$ to mean $m \mid (a - b)$. The number m is called the modulus.*

Given $m > 0$ and a, we can write $a = qm + b$ for a uniquely determined b with $0 \leq b < m$; clearly $q = [a/m]$ and $a \equiv b \pmod{m}$ here. This b is called the *least positive residue of* a mod m. (But *least nonnegative residue* might be better, for it *can* be 0.) When we need a notation we shall use $\langle a \rangle_m$ for this least positive residue. In PASCAL this least positive residue is obtained by $a - m * \mathrm{INT}(a/m)$ *provided* $a > 0$. The function INT 'rounds towards zero' in PASCAL so in fact

$[x] = \mathrm{INT}(x)$ if $x > 0$ or $x = \mathrm{INT}(x)$; otherwise $[x] = \mathrm{INT}(x) - 1$.

The use of INT produces an answer of type real (or extended), and it is usually best to do calculations using these types, for maximum precision.

In what follows, we shall always assume that a number occurring as a modulus is positive.

1.2 Properties of $\equiv$

(a) $a \equiv b \pmod{m}$ is *always* true if $m = 1$.

(b) If $a \equiv b$ and $b \equiv c \pmod{m}$, then $a \equiv c \pmod{m}$. [For $a - b = km$, $b - c = jm$ imply $a - c = (k + j)m$.]

(c) If $a \equiv b$ and $c \equiv d \pmod{m}$ then $a + c \equiv b + d$ and $a - c \equiv b - d$ $\pmod{m}$. [For $a - b = km$, $c - d = jm$ imply $(a + c) - (b + d) = (k + j)m$; similarly for the other statement.]

(d) If $a \equiv b \pmod{m}$ then, for any (integer) c, $ac \equiv bc \pmod{m}$. [For $a - b = km$ implies $ac - bc = kcm$.]

(e) If $a \equiv b$ and $c \equiv d \pmod{m}$, then $ac \equiv bd \pmod{m}$. [For $ac \equiv bc$ and $bc \equiv bd$, using (d). Now use (b).] Note the corollary: if $a \equiv b \pmod{m}$ then $a^k \equiv b^k \pmod{m}$ for any integer $k \geq 1$.

(f) $a \equiv b \pmod{m}$ if and only if $ac \equiv bc \pmod{mc}$. [This is immediate from the definition of $\equiv$.]

The next three properties are slightly more substantial, and we separate them off as a proposition.

1.3 Proposition

(a) *If $ac \equiv bc \pmod{m}$ and $(c, m) = d$, then $a \equiv b \pmod{m/d}$. In particular if c and m are coprime (i.e., $d = 1$), then $a \equiv b \pmod{m}$: c can be cancelled.*

(b) *If $a \equiv b \pmod{m}$ and $a \equiv b \pmod{n}$, then $a \equiv b \pmod{[m, n]}$, where $[m, n]$ is the lcm of m and n. In particular if m and n are coprime, then $a \equiv b \pmod{mn}$. (Compare 2.3(a) and 2.5 in Chapter 1.)*

(c) *If $a \equiv b \pmod{m_i}$ for $i = 1, \ldots, r$, where $(m_i, m_j) = 1$ for $i \neq j$, then $a \equiv b \pmod{m_1 m_2 \ldots m_r}$.*

Proof (a) Let $ac - bc = km$, and write $c = c_1 d$, $m = m_1 d$ where c_1 and m_1 are coprime (compare 2.2(b) of Chapter 1). Then $m_1 \mid (a - b)c_1$ and so, m_1 and c_1 being coprime, $m_1 \mid (a - b)$. This proves the result.

(b) If $a - b$ is a multiple of both m and n, then it is a multiple of $[m, n]$: compare the preamble to 2.5 of Chapter 1. Here, however, is a direct proof of the special case when m and n are coprime. Write $a - b = km = jn$. Then $m \mid jn$ and so, m and n being coprime, $m \mid j$. Writing $j = im$ gives $a - b = imn$, as required.

The proof of (c) is by repeated application of (b). $\square$

1.4 Exercises

(a) Give an example to show that, if $a \equiv b \pmod{m}$, then a^2 need not be congruent to $b^2 \pmod{m^2}$. (However the congruence *does* hold mod m, by 1.2(e).)

(b) Show that, for any integer a, $a^2 \equiv 0$ or $1 \pmod{4}$. Show also that

$a^4 \equiv 0$ or 1 (mod 16). [Write $a = 2k$ or $2k + 1$. In the latter case, it is helpful to remember that $k(k + 1)$ will always be even, since k or $k + 1$ will be even.]

(c) Deduce from the second part of (b) that, if a number of the form $16n + 15$ is expressed as the sum $a_1^4 + a_2^4 + \ldots + a_r^4$ of fourth powers of integers $a_i \geq 0$, then $r \geq 15$. (In fact it is known that every integer is expressible as a sum of 19 fourth powers. The number 79, for example, definitely needs 19 fourth powers: $79 = 4 \times 2^4 + 15 \times 1^4$ is the most economical expression. See for example [Ribenboim (1988), pp.236ff.] See also Chapter 1, 4.3(1) above.)

(d) Show that, for any a, $a^2 \equiv 0, 1$ or 4 (mod 8). Deduce that a number of the form $8k + 7$ cannot be written as the sum of three squares of integers ≥ 0.

(e) Show that, if n is odd, then $n^2 \equiv 1$ (mod 8). Show also that, if $3 \nmid n$, then $n^2 \equiv 1$ (mod 3). [Use $n \equiv 1$ or 2 (mod 3).] Deduce that, if n is odd and not a multiple of 3, then $n^2 \equiv 1$ (mod 24). (In particular this applies to any *prime* $n > 2$.) Why does it follow that, if n amd m are both odd and neither is a multiple of 3, then $24 \mid (n^2 - m^2)$?

(f) Show that, if p is prime, then $x^2 \equiv y^2$ (mod p) implies $x \equiv \pm y$ (mod p). [$p \mid (x - y)(x + y)$ and p is prime.] Give an example to show that this need not happen when p is composite.

(g) A particular case of (f) is $x^2 \equiv 1$ (mod p) implies $x \equiv \pm 1$ (mod p). Show that, if p is an *odd* prime and $k \geq 1$, then $x^2 \equiv 1$ (mod p^k) implies $x \equiv \pm 1$ (mod p^k). [The point here is that $p^k \mid (x - 1)(x + 1)$ and $(x - 1, x + 1) = 1$ or 2 so p cannot divide both factors. Compare Chapter 1, 2.2(d) above.]

(h) Show that, if p is prime and $k \geq 1$, then $x^2 \equiv x$ (mod p^k) implies that $x \equiv 0$ or $x \equiv 1$ (mod p^k). [The essential fact here is that $(x, x-1) = 1$.] Compare Chapter 1, 2.2(d).

(i) Let n be an integer ≥ 0 and let $x = [\sqrt{4n + 1}], y = [\sqrt{4n + 2}], z = [\sqrt{4n + 3}]$. Use the first part of (b) above to show that, of these three square roots, only the first could possibly be an integer. Show further that if k is an integer and $x \leq k \leq z$, then $k = x$. [You could consider $x \leq k < y$ and $y \leq k \leq z$ separately.] Deduce that $x = y = z$.

(j) Show that, for any integer n, $\sqrt{4n + 1} \leq \sqrt{n} + \sqrt{n + 1} \leq \sqrt{4n + 3}$. Deduce from (i) above that $x = y = z = [\sqrt{n} + \sqrt{n + 1}]$.

(k) Let r_n denote the 'repunit' $(10^n - 1)/9$, whose decimal representation is n ones. Show that $d \mid r_n$ if and only if $10^n \equiv 1$ (mod $9d$). Show that, if $d \mid r_a$ and $d \mid r_b$ then $d \mid r_{a+b}$ and, if $a > b$, then $d \mid r_{a-b}$.

(l) Let $n = a_0 + 10a_1 + 10^2a_2 + \ldots + 10^ka_k$, where $0 \leq a_i \leq 9$ for each i, so that the decimal representation of n is $a_ka_{k-1}\ldots a_1a_0$. Show that, modulo 3 or modulo 9, $n \equiv a_k + a_{k-1} + \ldots + a_1 + a_0$. Deduce the well-known test for divisibility by 3 or 9: a number is divisible by 3 (resp. by 9) if and only if the sum of its decimal digits is divisible by 3 (resp. by 9).

(m) Similarly show that a number is divisible by 11 if and only if the *alternating sum* $a_k - a_{k-1} + a_{k-2} - \ldots \pm a_0$ is divisible by 11.

(n) Suppose that a, b, c, d, x, y are integers satisfying $(ad - bc, m) = 1$ and $ax + by \equiv 0 \pmod{m}$, $cx + dy \equiv 0 \pmod{m}$. Show that $x \equiv 0$ and $y \equiv 0 \pmod{m}$. [Just as in ordinary linear equations, eliminate y by multiplying the first congruence by d and the second by b and subtracting.]

(o) Suppose that p is prime and $a > 1, m \geq 1$ are integers such that $a^{2^m} \equiv -1 \pmod{p}$. Show that, if $n > m$, then $a^{2^n} \equiv 1 \pmod{p}$. Hence show that the greatest common divisor of $a^{2^m} + 1$ and $a^{2^n} + 1$ is 1 if a is even and 2 if a is odd. [Hint. Show that, if p divides both numbers, then $p \mid 2$.]

(p) Show that, for any integers h and k,

$$h^2 + k^2 \equiv \begin{cases} 0, 1 \text{ or } 4 \pmod{8} & \text{if } h \equiv 0 \pmod{4}, \\ 1, 2 \text{ or } 5 \pmod{8} & \text{if } h \equiv 1 \text{ or } 3 \pmod{4}, \\ 0, 4 \text{ or } 5 \pmod{8} & \text{if } h \equiv 2 \pmod{4}. \end{cases}$$

Deduce that no solutions exist to the equations

$$(x + 1)^2 + a^2 = (x + 2)^2 + b^2 = (x + 3)^2 + c^2 = (x + 4)^2 + d^2,$$

for integers x, a, b, c, d.

(q) For an integer n, can all the numbers $n + 1$, $n + 3$, $n + 7$, $n + 9$, $n + 13$, $n + 15$ be prime? [Work modulo 5.]

(r) In the notation of Chapter 1,3.13(j), suppose that s is a prime and that $s \mid (n^{p-1} - 1)$ but $s \nmid (n^q - 1)$. Show that $s \mid (N - 1)$. Deduce that $2^{16} - 1$ divides $\frac{2^{85}-1}{(2^{17}-1)(2^5-1)} - 1$. (Note that $2^{16} - 1$ is *not* prime!)

(s) Show that, for n odd, $1 + 2 + 3 + \ldots + n \equiv 0 \pmod{n}$. Deduce that, when n is odd, it is not possible to find a permutation $a_0, a_1, \ldots, a_{n-1}$ of $\{0, 1, \ldots, n-1\}$ such that $a_1 - a_0, a_2 - a_1, \ldots, a_{n-1} - a_{n-2}$ are all distinct (and of course nonzero) mod n. For n even can you find such a permutation? (Such a permutation is sometimes called a *directed terrace*.)

(t) Show that, if $x \equiv \pm 1 \pmod{3}$, then $x^3 \equiv \pm 1 \pmod{9}$. Use this to show that it is not possible to find integers x, y, z, none of which is divisible by 3 and such that $x^3 + y^3 = z^3$. This is the 'first' or 'easy' case of *Fermat's last theorem* for exponent 3. It is much harder to prove that the equation has no solutions at all in nonzero integers. See for example [Edwards (1977)].

(u) Show that, if $n \geq 5$, then $2^{n!} \equiv 76 \pmod{100}$.

The following result looks very innocent but in fact it will play a major role in the factorization method of Chapter 10.

1.5 Proposition *Suppose that integers x, y, n satisfy $x^2 \equiv y^2$ (mod n), but that $x \not\equiv y, x \not\equiv -y$ (mod n). Then $(x - y, n)$ is a proper factor of n (i.e., a factor other than 1 or n). The same applies to $(x + y, n)$.*

Proof Of course $h = (x - y, n)$ is a factor of n. Now $n \mid (x - y)(x + y)$. If $h = 1$, then it follows that $n \mid (x + y)$, that is, $x \equiv -y \pmod{n}$, contrary to hypothesis. If $h = n$, then $n \mid (x - y)$ (since $h \mid (x - y)$), that is, $x \equiv y \pmod{n}$, again contrary to hypothesis. So h is a proper factor of n. The proof for $(x + y, n)$ is similar.   □

1.6 Covering congruences Suppose that *every* positive integer n satisfies at least one of the congruences

$$x \equiv a_1 \pmod{n_1}, \quad x \equiv a_2 \pmod{n_2}, \quad x \equiv a_k \pmod{n_k},$$

where we suppose all the moduli $n_1, \ldots, n_k$ are distinct and > 1. Then these congruences are said to form a *covering set of congruences*. For example, $x \equiv 0 \pmod 2$, $x \equiv 0 \pmod 3$, $x \equiv 1 \pmod 4$, $x \equiv 5 \pmod 6$, $x \equiv 7 \pmod{12}$ is such a set. To see this, work modulo 12: the first congruence covers all numbers $\equiv$ 0, 2, 4, 6, 8, 10; the second all numbers $\equiv$ 0, 3, 6, 9; the third all numbers $\equiv$ 1, 5, 9; the fourth all numbers $\equiv$ 5, 11; and the fifth all numbers $\equiv$ 7. Thus all numbers are covered, but notice that none of the congruences is redundant. Clearly the least common multiple of the moduli (in this case the largest modulus) plays a role in the study of covering congruences. We conclude this section with some results and problems concerning covering congruences.

1.7 Proposition *If the congruences $x \equiv a_i \pmod{n_i}$, $i = 1, \ldots, k$, are covering congruences in the sense of 1.6, then $\sum \frac{1}{n_i} \geq 1$.*

Proof Let $N = \operatorname{lcm}(n_1, \ldots, n_k)$ and let $N_i = N/n_i$. Then, mod N, the solutions of the ith congruence are $x \equiv a_i, a_i + n_i, \ldots, a_i + (N_i - 1)n_i$, which is N_i numbers in all. Taking $i = 1, \ldots, k$ in turn all the N numbers mod N must be covered, so $\sum N_i \geq N$, which is equivalent to the result.   □

One way to look for sets of covering congruences is as follows. Choose an integer N, such as 120, with plenty of divisors. All the moduli n_i will be factors of N. Consider for example $n_1 = 3$: the congruence $x \equiv 1$ (mod 3) covers 40 numbers (mod 120) and that is as many as is possible for this n_1. Now choose the next divisor $n_2 = 4$. You can verify that $x \equiv 1 \pmod 4$ covers an additional 20 numbers (mod 120) and that is as many as possible for this n_2. Now go on, choosing for each successive divisor n_i an optimum a_i which covers as many extra numbers (mod 120) as possible. With luck (which holds good for 120) you will find a set of covering congruences.

1.8 Project Carry out the above method and so produce a set of covering congruences all to moduli which are factors of 120. Notice that the smallest modulus is 3. There exist sets of covering congruences with smallest modulus 4: the above method will produce one with $N = 720$ but there also exists one with $N = 360$. There is some information on this in [Churchhouse (1968)].

1.9 Exercise Deduce from 1.7 above that no set of covering congruences can exist which is based, in the above way, on the divisors of a number of the form $3^a 5^b$. [Hint. Work out $\sum\sum 3^{-r}4^{-s}$ where r and s run from 0 to ∞. Then subtract 1 to disallow $r = s = 0$; you should find the answer is < 1.]

1.10 Computing exercise: Fibonacci sequences mod m Let $m \geq 2$ and $f_0 = 0$, $f_1 = 1$, $f_{n+2} \equiv f_{n+1} + f_n \pmod m$ for $n \geq 0$. The sequence of f_i will repeat when two consecutive terms are repeated as consecutive terms later in the sequence. Why does this imply that the sequence necessarily repeats? Show that if $f_n \equiv f_{n+k}$, $f_{n+1} \equiv f_{n+k+1}$, where $k > 0$, then in fact $f_0 \equiv f_k$, $f_1 \equiv f_{k+1}$ (all mod m) so that the sequence repeats from the beginning. Taking $k = k(m)$ as small as possible we call $k(m)$ the *period of the Fibonacci sequence* mod m. (Compare Chapter 1, 5.4.) For example with $m = 7$ the terms are (mod 7) 0,1,1,2,3,5,1,6,0,6,6,5,4,2,6,1 and then it repeats, so $k(7) = 16$. Write a program to calculate $k(m)$ from m, and verify the following:

(a) If m is a prime $\equiv \pm 1 \pmod{10}$, then $k(m) \mid (m - 1)$. (We shall take this up again in Chapter 11, 5.1(b).) For which values of $m < 1000$ is $k(m)$ a *proper* divisor of $m - 1$?

(b) If m is a prime $\equiv \pm 3 \pmod{10}$, then $k(m) \mid (2m + 2)$. For which values of $m < 1000$ is $k(m)$ a *proper* divisor of $2m + 2$?(For $m = 7, k(m) = 2m + 2$.)

(c) For any prime $p < 1000$, $k(p) \neq k(p^2)$.

For further information see [Wall (1960), Ehrlich (1989)].

2. Inverses mod m and solutions of certain congruences

In this section we answer the question: given a, when does there exist b

such that $ab \equiv 1 \pmod{m}$? We also calculate b from a when b exists; b is called the *inverse* of $a \pmod{m}$ and, with due care, we can write $b \equiv a^{-1}$.

2.1 Proposition *Given a (and m), there exists b such that $ab \equiv 1$ (mod m) if and only if $(a,m) = 1$. When b exists, it is also coprime to m, and is unique (mod m). The last statement means: if $ab \equiv ab' \equiv 1$ (mod m), then $b \equiv b'$ (mod m).*

Proof If $ab \equiv 1 \pmod{m}$, then $ab = 1 + km$, so that any common factor of a and m must divide 1, which gives $(a,m) = 1$ (also $(b,m) = 1$). Conversely, if $(a,m) = 1$ then, by Euclid's algorithm, there exist integers b and k such that $ab + mk = 1$ (compare Chapter 1,3.12). For this b, $ab \equiv 1 \pmod{m}$. The uniqueness of b follows from 1.3 (a). $\square$

Note the corollary of 2.1: *if p is prime, then every number which is not a multiple of p has an inverse* (mod p). (This is expressed in group-theoretic terms by saying that the numbers $\{1, 2, \ldots, p-1\}$ form a group under multiplication mod p.)

Once we can find inverses we can solve equations $ax \equiv c \pmod{m}$ for the unknown x. (Compare Chapter 1,3.14 and 3.17, on linear Diophantine equations.) Let $(a,m) = h$. Clearly, if $h \nmid c$, then the congruence is not solvable for x (for $ax = c + km$ implies $h \mid c$). So assume $h \mid c$ and write $a = a'h$, $m = m'h$, $c = c'h$, so that $(a',m') = 1$. Dividing $ax = c + km$ through by h, we obtain $a'x \equiv c' \pmod{m'}$. To solve this, multiply both sides by b', where $a'b' \equiv 1 \pmod{m'}$: we obtain $x \equiv b'c' \pmod{m'}$. Thus there is a unique solution $x = x_1$, mod m', to $ax \equiv c \pmod{m}$. The solutions x, mod m, are just $x_1, x_1 + m', x_1 + 2m', \ldots, x_1 + (h-1)m'$. $\square$

2.2 Exercises

(a) Solve the following congruences (or show that they have no solution) using the above method:

$$2x \equiv 3 \pmod{7}; 10x \equiv 3 \pmod{18}; 2x \equiv 3 \pmod{15};$$

$$10x \equiv 15 \pmod{17}; 10x \equiv 3 \pmod{27};$$

$$35x \equiv 14 \pmod{182}.$$

[Answers to the last two: $x \equiv 3 \pmod{27}$; $x \equiv 16 \pmod{26}$.]

(b) Suppose that $ax \equiv c \pmod{m}$, $0 < a < m$. Let $m = ka + a_1$, where $0 \le a_1 < a$; thus a_1 is the least positive residue of $a \pmod{m}$ and $k = [m/a]$. Use $axk \equiv ck \pmod{m}$ to show that $a_1 x \equiv -ck \pmod{m}$, which is another congruence of the same form as the one we started with but with a coefficient of x which is *smaller* (and ≥ 0). Repeating the process, the coefficient of x must eventually become 0; hence at the previous step, the coefficient a' of x divides m, i.e., the congruence has the form $a'x \equiv c' \pmod{m}$. Of course if $a' \nmid c'$ then this has no solutions; otherwise, it is equivalent to $x \equiv (c'/a') \pmod{(m'/a')}$. Quite often, the coefficient of x becomes 1 on the step before it becomes 0; in any case solutions of the final congruence can be read off. The solutions of the original congruence are among the latter, but some solutions of the final congruence may fail to satisfy the original one.

For example, consider $17x \equiv 13 \pmod{122}$, which has a unique solution $\pmod{122}$ since 17 and 122 are coprime. Successive reductions are $3x \equiv 31 \pmod{22}$ and $2x \equiv 102 \pmod{122}$. The latter is equivalent to $x \equiv 51 \pmod{56}$, which has *two* solutions mod 122: $x \equiv 51$ and $x \equiv 107$, but only $x \equiv 51 \pmod{122}$ satisfies the original congruence.

Solve the congruences in (a) above by this method. Note that it is possible for a congruence with *no* solutions to yield a congruence with solutions!

2.3 Computing exercises

(a) (Compare Chapter 1, 3.17). Use the program 3.16 of Chapter 1 to find solutions to congruences $ax \equiv c \pmod{m}$ where a and m are coprime. Extend it to find solutions when $(a, m) = h$ and $h \mid c$. Check that the program is working by solving again the congruences in 2.2(a) above.

(b) Write a program to solve congruences $ax \equiv c \pmod{m}$ where $(a, m) = 1$, using the method of 2.2(b) above.

2.4 Exercises In this exercise, we shall find the solutions to the congruence $x^2 \equiv 1 \pmod{pq}$, where p and q are distinct primes.

(a) Show that $x^2 \equiv 1 \pmod{pq}$ if and only if $x^2 \equiv 1 \pmod{p}$ and $x^2 \equiv 1 \pmod{q}$.

(b) Consider $x^2 \equiv 1 \pmod{p}$, which has solutions $x \equiv \pm 1 \pmod{p}$ (compare 1.4(f)). Write $x = \pm 1 + kp$; show that this satisfies $x^2 \equiv 1 \pmod{q}$ if and only if (i) $k \equiv 0 \pmod{q}$, giving $x \equiv \pm 1 \pmod{pq}$, or (ii) $kp \pm 2 \equiv 0 \pmod{q}$.

(c) Now let r satisfy $pr \equiv 1 \pmod{q}$. Show that the solutions to $x^2 \equiv 1 \pmod{pq}$ coming from (ii) above are $x = \pm(1 - 2pr) \pmod{pq}$.

(d) Find the four solutions to $x^2 \equiv 1 \pmod{pq}$ where $p = 101$ and $q = 113$.

(e) Show that, if p and q are both odd, then the four solutions $\pm 1, \pm(1 - 2pr)$ are all distinct $\pmod{pq}$, whereas, if p or q is 2, then the congruence $x^2 \equiv 1 \pmod{pq}$ has just two solutions.

2.5 Exercise There is an interesting addition to 2.4 when p and q are both odd. In fact, let x_1 be a solution to the congruence $x^2 \equiv 1$ (mod pq), where $x_1 \not\equiv \pm 1$. Using 1.5 we find that $(x_1 + 1, pq)$ is a proper factor of pq, and hence equals p or q. The same goes for $(x_1 - 1, pq)$. Check these explicitly using the formula in 2.4(c) above for the solutions.

It is quite easy to solve simultaneous linear congruences by successive substitution. Here is an example.

Find all x such that $6x \equiv 8$ (mod 10) and $4x \equiv 7$ (mod 9).

Now $6x \equiv 8$ (mod 10) if and only if $3x \equiv 4$ (mod 5). Since $3 \cdot 2 \equiv 1$ (mod 5), this has solutions $x \equiv 8 \equiv 3$ (mod 5). Write $x = 3 + 5k$. Then $4x \equiv 7$ (mod 9) gives $2k \equiv 4$ (mod 9), i.e., $k \equiv 2$ (mod 9). Hence $k = 2 + 9j$ and

$$x = 3 + 5(2 + 9j) = 13 + 45j.$$

It follows that the general solution is $x \equiv 13$ (mod 45).

2.6 Exercises (a) Solve the following sets of simultaneous congruences.

$3x \equiv 3$ (mod 15) and $4x \equiv 2$ (mod 21).   [Answer: $x \equiv 11$(mod 105).]

$10x \equiv 10$ (mod 25) and $3x \equiv 8$ (mod 10) and $7x \equiv 12$ (mod 15).

[Answer: $x \equiv 6$ (mod 30).]

(b) What is the smallest number greater than 1 which leaves remainder 1 when divided by all the numbers $2, 3, \ldots, 10$? [What is the least common multiple of $2, 3, \ldots, 10$?] What is the smallest positive number which leaves remainder 1 when divided by 2, remainder 2 when divided by 3, and so on, up to remainder 9 when divided by 10? [Why is this equivalent to $x \equiv -1$ (mod lcm$(2, 3, \ldots, 10)$)?]

(c) Approximately how big is the smallest number greater than 1 which leaves remainder 1 on division by all the numbers $2, 3, \ldots, 50$? [Answer: about 3.1×10^{21}.]

(d) The least common multiple of the numbers $1, 2, \ldots, 12$ is 27 720. If x is to leave remainder $i - 1$ when divided by i, for each $i = 2, 3, \ldots, 12$, then it follows that $x \equiv -1$ (mod 27 720). What is the smallest such number that is exactly divisible by 13?

(e) Suppose that x is an integer, $0 \le x \le 99$, and the final two digits of x^2 coincide with the digits of x. The object is to find the possible values of x. Note that this is equivalent to solving $x^2 \equiv x$ (mod 100), which is equivalent to $x^2 \equiv x$ (mod 4) and $x^2 \equiv x$ (mod 25) (see 1.3(b)). Using 1.4(h), show that there are four cases: $x \equiv 0$ (mod 4 and mod 25); $x \equiv 1$ (mod 4 and mod 25); $x \equiv 0$ (mod 4) and $x \equiv 1$ (mod 25); $x \equiv 1$ (mod 4) and $x \equiv 0$ (mod 25). Now solve these to find all the solutions for x

You can find many examples of congruences, some of them very hard to solve because they involve decimal expansions of numbers, in [Beiler(1966), Chapters V and XXV]. For example [p.300]: what is the smallest number consisting only of digits 3, 5 and 7 such that both the number and the sum of the digits are divisible by 3, 5 and 7? [Answer: 33 577 577 777 777 775.]

2.7 Computing exercise   Write a program to solve the simultaneous congruences $ax \equiv b \pmod{m}$ and $cx \equiv d \pmod{n}$, assuming that (a, m) and (c, n) are both 1.

2.8 Computing exercises   These could be done by hand, but you may find the program in 2.7 (or that in 2.3(a) together with some hand calculation) useful to handle the numbers involved.

(a) Find all numbers x with $0 \leq x \leq 999$ satisfying $x^2 \equiv x \pmod{1000}$. Compare 2.6(e) above. We are looking for numbers which coincide with the last three digits of their squares.

(b) The object here is to solve $3x^2 \equiv x \pmod{10^4}$. First check that $(x, 3x - 1) = 1$, and then reduce to four cases as in 2.6(e): $x \equiv 0 \pmod{2^4}$ and mod 5^4); $3x \equiv 1 \pmod{2^4}$ and mod 5^4); $x \equiv 0 \pmod{2^4}$ and $3x \equiv 1 \pmod{5^4}$); $3x \equiv 1 \pmod{2^4}$ and $x \equiv 0 \pmod{5^4}$.

(c)* Solve the congruence $x^3 \equiv x \pmod{1000}$ for $0 \leq x \leq 999$. This is slightly more tricky than (a) and (b) since $x^3 - x = x(x - 1)(x + 1)$ and these three factors are not pairwise coprime when x is odd.

There is an interesting result about congruences $ax \equiv y$ modulo a *prime p*, known as Thue's lemma (1902). We shall have occasion to use it later.

2.9 Thue's lemma   *Let p be prime and $p \nmid a$. Then there exist x and y with $0 <| x |< \sqrt{p}$ and $0 <| y |< \sqrt{p}$ satisfying $ax \equiv y \pmod{p}$.*

Proof   This is proved by a counting argument. Consider numbers $au - v$ where $u = 0, 1, \ldots, [\sqrt{p}]$ and $v = 0, 1, \ldots, [\sqrt{p}]$. There are clearly $([\sqrt{p}] + 1)^2$ such numbers. Now $[\sqrt{p}] + 1 > \sqrt{p}$, so $([\sqrt{p}] + 1)^2 > p$, and consequently the numbers $au - v$ cannot all be different modulo p: there must exist u_1, v_1, u_2, v_2 such that $u_1 \neq u_2$ or $v_1 \neq v_2$ and $au_1 - v_1 \equiv au_2 - v_2 \pmod{p}$. This last condition is the same as $a(u_1 - u_2) \equiv v_1 - v_2 \pmod{p}$, and from this it follows that *neither* $x = u_1 - u_2$ nor $y = v_1 - v_2$ can be zero (for either zero implies the other zero, as $p \nmid a$). Hence these x and y satisfy the conclusion of the lemma.   □

We conclude this section with a fundamental theorem on solution of simultaneous linear congruences in which the moduli are coprime in pairs, together with a natural generalization to general moduli. (The theorem takes its name from its origins in ancient China: it appears in Ch'in Chiu-shao's *Shu-shu Chiu-chang* (Mathematical treatise in nine sections) of 1247.)

2.10 Chinese remainder theorem *Let $m_1, m_2, \ldots, m_r$ be positive, and with every pair coprime: $(m_i, m_j) = 1$ when $i \neq j$. Then the simultaneous congruences*

$$x \equiv a_1 (\bmod\ m_1), x \equiv a_2 (\bmod\ m_2), \ldots, x \equiv a_r (\bmod\ m_r)$$

have a unique solution modulo the product $m_1 m_2 \ldots m_r$.

Proof The uniqueness follows from 1.3(c). The existence of a solution can be proved by induction on r, or by the following explicit construction. Let k_i, for $i = 1, \ldots, r$, be the product of the ms but with m_i omitted. Certainly $(k_i, m_i) = 1$, so k_i has an inverse n_i modulo m_i. Let

$$x = a_1 k_1 n_1 + a_2 k_2 n_2 + \ldots + a_r k_r n_r.$$

This is the required solution. For instance, m_1 is a factor of $n_2, \ldots, n_r$ and $k_1 n_1 \equiv 1 \pmod{m_1}$. Hence $x \equiv a_1 \pmod{m_1}$. Similarly for the other congruences to be satisfied.

2.11 Exercises

(a) Solve the congruences $x \equiv 4 \pmod 7$, $x \equiv 6 \pmod 9$, $x \equiv 1 \pmod{10}$, both by successive substitution as in the example before 2.6 and by the explicit formula in 2.10. [Answer: $x \equiv 501 \pmod{630}$.] Of course, you can make up your own congruences to order, by choosing pairwise coprime moduli and starting with the answer!

(b) Suppose that p is a prime and there exists a with $a^2 \equiv -1 \pmod p$. (We shall see later that this is true precisely for primes of the form $4k+1$. See Chapter 11,1.5(a).) Use Thue's lemma 2.9 to deduce that there exist x and y with $x^2 + y^2 \equiv 0 \pmod p$ and $|x|, |y|$ strictly between 0 and $\sqrt{p}$. Deduce that in fact $x^2 + y^2 = p$. (So p is the sum of two squares.)

(c) Let m and n be coprime integers with m odd. Verify using the Chinese remainder theorem that we can find integers $a > 0$, $b > 0$ with m a factor of both $2a + 1$ and $2b$, and n a factor of both a and $b + 1$. Now consider the equation $X^2 + Y = Z^2$, which certainly has solutions in integers X, Y, Z all > 0. Multiply through by $Y^{2a} Z^{2b}$ and hence find integer solutions to $x^2 + y^m = z^{2n}$. Is there a corresponding result for m even? (Compare [Boyd (1990)].)

2.12 Project: expressing a prime of the form $4k+1$ as the sum of two squares There is a wonderful algorithm for finding integers x and y such that $x^2 + y^2 = p$, where p is a given prime of the form $4k + 1$. We merely state it here and suggest that you implement it in a computer program; the details can be found in [Wagon (1990)] and require a small acquaintance with quadratic residues (see Chapter 11, but the algorithm can be written without such acquaintance). We need to find a as in 2.11(b) above with $a^2 \equiv -1 \pmod p$. Now, for any c with $p \nmid c$, we have $c^{(p-1)/2} \equiv \pm 1 \pmod p$; by trial starting with $c = 2$ we find a c which gives $c^{(p-1)/2} \equiv -1 \pmod p$. (This amounts to saying that c is *not a perfect square* mod p; compare 2.5 of Chapter 8.) We then take $a = c^{(p-1)/4}$, recalling that $p - 1$ is divisible by 4.

The wonderful algorithm is then as follows: Apply Euclid's algorithm (see 3.5 of Chapter 1 for the program) to p and a; the first two remainders that are less than $\sqrt{p}$ are suitable integers x and y with $x^2 + y^2 = p$. (Note that the remainders eventually reach 1, so there will certainly be remainders less than $\sqrt{p}$.)

It is not hard to extend the Chinese remainder theorem to general moduli, though there is not such a pleasantly simple formula for the solution. We have already met some examples of this in 2.6 above. The result is as follows:

2.13* Simultaneous linear congruences with arbitrary moduli
The congruences $x \equiv a_i \pmod{m_i}$, $i = 1, \ldots, r$, have a simultaneous solution if and only if, for all $i \neq j$, $(m_i, m_j) \mid (a_i - a_j)$. The solution is then unique mod $[m_1, \ldots, m_r]$ = the least common multiple of $m_1, \ldots, m_r$.

Proof If $x \equiv a_i \pmod{m_i}$ and $x \equiv a_j \pmod{m_j}$ then $a_i + \lambda m_i = a_j + \mu m_j$ for some integers λ and μ, from which $(m_i, m_j) \mid (a_i - a_j)$. If x and y are both solutions of all the congruences, then $x \equiv y \pmod{m_i}$ for all i, which holds if and only if $x \equiv y \pmod{[m_1, \ldots, m_r]}$.

It remains to show that the given conditions are *sufficient* for a solution, so suppose $(m_i, m_j) \mid (a_i - a_j)$ for all $i \neq j$. Consider first the special case where $m_i = p^{k_i} (i = 1, \ldots, r)$, where p is prime, and we may suppose that $k_1 \geq k_2 \geq \ldots \geq k_r$. Then $x \equiv a_1 \pmod{p^{k_1}}$ satisfies all the congruences: for any λ, and $i \geq 1$, $a_1 + \lambda p^{k_1} \equiv a_i \pmod{p^{k_i}}$ since $p^{k_i} = (m_1, m_i) \mid a_1 - a_i$.

Now consider the general case. Write $m_i = p_1^{k_{i1}} \ldots p_s^{k_{is}}$ for $i = 1, \ldots, r$, where the k_{ij} are ≥ 0 and $p_1, \ldots, p_s$ are distinct primes. Then the given congruences are equivalent to the rs congruences $x \equiv a_i \pmod{p_j^{k_{ij}}}$, $i = 1, \ldots, r$; $j = 1, \ldots, s$. For each prime p_j select the one of these congruences with the *largest* power of p_j, say $x \equiv b_j \pmod{p_j^{n_j}}$, where $n_j = \max(k_{1j}, \ldots, k_{rj}) = k_{ij}$ say, and $b_j = a_i$ for this i which gives the maximum k_{ij}. Then by the special case already considered, all the congruences involving p_j are satisfied by $x \equiv b_j \pmod{p_j^{n_j}}$. These s

congruences are to coprime moduli and therefore have a solution by the Chinese remainder theorem. (And of course the product of these $p_j^{n_j}$ is precisely the least common multiple of $m_1, \ldots, m_r$.) $\qquad\square$

As an example, consider $x \equiv 5$ (mod 6), $x \equiv 9$ (mod 10) and $x \equiv 2$ (mod 9). The hypotheses of 2.13 are satisfied, and we solve by successive substitution. Thus $x = 5 + 6u$, substituted in the second congruence, gives $6u \equiv 4$ (mod 10), i.e. $3u \equiv 2$ (mod 5), so that $u \equiv 4$ (mod 5), $u = 4 + 5v$ say, so that $x = 29 + 30v$. Substituting in the third congruence gives $30v \equiv 0$ (mod 9), i.e. $10v \equiv 0$ (mod 3), so that $v \equiv 0$ (mod 3) and finally $x \equiv 29$ (mod 90) is the solution. Note that the reduction given in the proof of 2.13 turns the given congruences into $x \equiv 1$ (mod 2), $x \equiv 4$ (mod 5) and $x \equiv 2$ (mod 9), which could be solved by the formula of the proof of the Chinese remainder theorem 2.10.

2.14* Project Write a program to solve any three simultaneous congruences $x \equiv a$ (mod m), $x \equiv b$ (mod n), $x \equiv c$ (mod k), first checking whether there is a solution using the criterion of 2.13. (There is a formula for the lcm of three numbers in Chapter 1, 2.6(e), if you feel the need of it.) Check that the solution to $x \equiv 20$ (mod 28), $x \equiv 10$ (mod 90), $x \equiv 55$ (mod 105) is $x \equiv 1000$ (mod 1260), and make up your own examples to test the program further.

3. Further examples of congruences

3.1 Project: primes in arithmetic progression Let $r \geq 3$ and let d be even and ≥ 4. We want to find r consecutive primes in arithmetic progression:
$$p, p + d, p + 2d, \ldots, p + (r - 1)d.$$
(a) Let $r = 3, d = 6$. Thus, $p, p + 6, p + 12$ are to be consecutive primes. Show that, if $3 \mid (p + 1)$, $5 \mid (p + 3)$ and $7 \mid (p + 2)$ then $p + i$ is composite for $1 \leq i \leq 11, i \neq 6$. Show that these divisibility conditions hold provided $p \equiv 47$ (mod 105). Write a program to determine which $p < 10\,000$ and satisfying this congruence produce the required three consecutive primes in arithmetic progression.

(b) Assuming $p > 3$ and $p, p + d, p + 2d$ are all primes (so $r = 3$ again), show that $d \equiv 0$ (mod 3). (Hint: Show that, if $3 \nmid d$, then p, $p + d$, $p + 2d$ will all be different (mod 3). Why is this a contradiction?) Since d is even, this shows $d \equiv 0$ (mod 6). [There is a similar result for five primes (all > 5) in arithmetic progression with common difference d: we have $2 \cdot 3 \cdot 5 = 30 \mid d$. No doubt you can see the general result here! It seems to date back to Waring in 1770.]

(c) Let $r = 4, d = 6$. Show that if $3 \mid (p+1)$, $5 \mid (p+4)$, $7 \mid (p+2)$ and $11 \mid (p+8)$, then $p + i$ is composite for $1 \leq i \leq 17$, $i \neq 6, 12$. Show that these divisibility conditions hold for $p \equiv 971 \pmod{1155}$ and find the smallest such p giving four consecutive primes in arithmetic progression. (Hint: this prime is between $200\,000$ and $300\,000$.) You could also find the smallest prime p of all which makes p, $p+6$, $p+12$, $p+18$ consecutive primes.

3.2 Project: more primes in arithmetic progression

The three primes starting with 3, namely 3, 5, 7, are in arithmetic progression. Find five primes 5, $5 + d$, $5 + 2d$, $5 + 3d$, $5 + 4d$ and find seven primes 7, $7 + d, \ldots, 7 + 6d$ (this requires a different value of d). (Unfortunately the smallest d for which $11, 11 + d, \ldots, 11 + 10d$ are all prime is $d = 1\,536\,160\,080$.) Guided by the following hints, write out a proof of the theorem: if p, $p + d$, $p + 2d, \ldots, p + (p - 1)d$ are all prime, then d is divisible by *every* prime q with $2 \leq q < p$. (For example with $p = 7$, d must be divisible by 2, 3 and 5, and hence by 30.)

Assume, for a contradiction, that q is prime, $q < p$ and $q \nmid d$. Deduce that no two of the numbers p, $p+d, \ldots, p+(q-1)d$ are congruent modulo q. Since there are q of them, why does it follow that one of them is $\equiv 0 \pmod{q}$? Use $q \mid (p + jd)$ for some j, $0 \leq j \leq q - 1$, and the fact that q and $p + jd$ are both prime, to get a contradiction.

3.3* Exercise: solutions of $x^2 \equiv 1 \pmod{m}$

This exercise is really to pave the way for 3.4; it generalizes the results of 1.4(g) and 2.4. Let the modulus m be

$$m = 2^a p_1^{a_1} p_2^{a_2} \ldots p_r^{a_r},$$

where the p_i are the odd primes in ascending order. The idea is to prove that the number of solutions of $x^2 \equiv 1 \pmod{m}$ is

$$2^{r+e} \text{ where } e = \begin{cases} 0 & \text{if } a = 0 \text{ or } 1, \\ 1 & \text{if } a = 2, \\ 2 & \text{if } a > 2. \end{cases}$$

The point is that $x^2 \equiv 1 \pmod{m}$ if and only if $x^2 \equiv 1 \pmod{2^a}$ and $\pmod{p_1^{a_1}}$ and $\ldots$ and $\pmod{p_r^{a_r}}$ since these moduli are coprime in pairs. Furthermore, once you specify a solution of each of these $r + 1$ congruences, a unique solution of the original congruence $\pmod{m}$ is determined, by the Chinese remainder theorem. It isn't hard to determine the solutions of the congruence mod 2^a, and the other congruences have solutions ± 1, by 1.4(g).

In particular, show that for $a > 2$ the solutions of $x^2 \equiv 1 \pmod{2^a}$ are ± 1 and $\pm(1 + 2^{a-1})$, and that these four numbers are distinct $\pmod{2^a}$.

3.4* Exercise: a theorem of Gauss For a fixed number $m > 1$, consider the numbers x which are coprime to m and satisfy $1 \le x < m$. (We shall meet these numbers a good deal in what follows; sometimes they are called a reduced set of residues modulo m.) Gauss's theorem gives the *product* P of these numbers modulo m: it asserts that this product is $+1$, unless $m = 4$, or m has the form p^k or $2p^k$ for an odd prime p and $k \ge 1$, in which case the product is -1. The idea is very simple: each number x as above has an inverse y such that $xy \equiv 1$ (mod m), and furthermore there is a unique such y with $1 \le y < m$. If $y \ne x$, then, in the product P, x and y will both occur and contribute 1 when multiplied together. The condition for, on the contrary, $x = y$, is $x^2 \equiv 1$ (mod m), so P is the same as the product (mod m) of all the numbers x with $1 \le x < m$ and $x^2 \equiv 1$ (mod m).

Now use 3.3 to prove Gauss's theorem.

3.5 Project: magic squares (You can find an extensive discussion of this subject in [Stark (1970), Chapter 4].)

Consider an $n \times n$ array where the rows are labelled $0, 1, \ldots, n-1$ and similarly the columns are labelled $0, 1, \ldots, n-1$. We want to fill this array with the integers $0, 1, \ldots, n^2 - 1$. Here is a possible prescription for doing so: place the integer k in row i and column j where, modulo n,

$$i \equiv e + ak + b[k/n], \qquad j \equiv f + ck + d[k/n] \qquad (1)$$

and a, b, c, d, e, f are fixed integers. (The significance of having k and $[k/n]$ is that, if these are known modulo n, then k is known precisely since it lies between 0 and $n^2 - 1$.) For example with $n = 4$, $e = f = 0$, $a = d = 1$, $b = c = 0$ we get the left-hand array, and with $n = 3$, $e = f = 0$, $a = 2$, $b = c = d = 1$ we get the right-hand array.

0	4	8	12		0	8	4
1	5	9	13		7	3	2
2	6	10	14		5	1	6
3	7	11	15				

The right-hand array is called a *magic square* because all rows and columns add up to the same thing (namely 12).

(a) Assume that $(ad - bc, n) = 1$. Show that (1) will never place two integers k and k' in the same position in the array. [Use 1.4(n).] It follows, of course, that the array is exactly filled by the numbers $0, 1, 2, \ldots, n^2 - 1$.

(b) Now assume that, as well as $(ad - bc, n) = 1$, we have $(a, n) = (b, n) = (c, n) = (d, n) = 1$. Consider a particular row in the array (fixing i), so that $ak + b[k/n] \equiv u$ (mod n) for a fixed u and all k in the row. Show that two integers in different positions in this row cannot have the

same value of k (mod n), by showing that they would then have the same
value of $[k/n]$ (mod n) too (using $(b,n) = 1$) and so would be the same
integer, contrary to (a). Similarly show that two integers in different
positions in the row cannot have the same value of $[k/n]$ (mod n).

It now follows that, for any row, the n integers k in that row have different values for k (mod n) and for $[k/n]$ (mod n). Deduce from this that
the sum of the integers in the row is $\frac{1}{2}n(n^2 - 1)$: the same for each row.

Similarly deduce that all the column sums are the same, so that the
square is in fact magic.

(c) Use the above method to write a program for producing magic
squares of order 3 and 5 (ideally printing out the numbers in a square
array on the screen). What is wrong with $n = 4$?

3.6 Project: Pollard's ρ-method for factorization   *[J.M. Pollard (1975).]* We're not really in a position to understand why this method
might possibly work, but it makes a nice computer program and it is at
least easy to understand the method itself! There is an extensive discussion in [Riesel (1985),pp.174ff]; see also [Niven *et al.* (1991), pp.80ff].
Let $f(x) = x^2 + 1$ and write $f^2(x) = f(f(x)), f^3(x) = f(f^2(x))$, etc.
Let $a_k = f^k(1)$. (Thus $a_1 = 2$, $a_2 = 5$, $a_{k+1} = a_k^2 + 1$.) Of course, other
polynomials could be used instead of this f. Suppose that we wish to
factorize a number N, that is to find a prime p (possibly the smallest
prime) with $p \mid N$. Of course the sequence $\{a_k\}$, taken mod p or mod
N, will eventually repeat and it is to be expected that it repeats much
sooner mod p than mod N. (In fact repetition becomes a reasonable
possibility once the number of terms exceeds $\sqrt{p}$.) Thus we hope that
$a_i \equiv a_j$ (mod p) for $j > i$ and j not too large. Note that then $a_t \equiv a_{j-i+t}$
(mod p) for all $t \geq i$. Of course we can't work out a_i mod p since we
don't know what p is, but we *can* work out a_i mod N, and we clearly
have

$$p \mid ((a_j - a_i) \bmod N, N).$$

(The reason for not working with $a_j - a_i$ directly is that it will rapidly
get *enormous*.)

There is one easy refinement: since $i < j$, there is an integer t with
$i < t \leq j$ and $(j - i) \mid t$. (This becomes clear if you ask why there must
be an integer k with $i/(j - i) < k \leq j/(j - i)$ and then use $t = k(j - i)$.)
Furthermore $a_t = a_{k(j-i)} \equiv a_{(k+1)(j-i)} \equiv a_{(k+2)(j-i)} \equiv \ldots \equiv a_{2k(j-i)} = a_{2t}$ (mod p) so instead of looking for i and j we might as well just
consider a_t and a_{2t}, for $t = 1, 2, 3, \ldots$. Thus the method is as follows:

Put $x = a_1 = 2, y = a_2 = 5$.

Replace x by $f(x), y$ by $f(f(y))$; thus now $x = a_2, y = a_4$. Find $(x - y, N)$.

Repeat: now $x = a_3, y = a_6$. Find $(x - y, N)$.

Keep going in this way until $(x - y, N)$ is neither 1 nor N (or until so many steps have been completed that you're tired of waiting). Of course x and y should be reduced modulo n at every stage, using $x := x - N*\text{INT}(x/N)$, etc. Note that the above method is *much* better than storing the a_i in an array; even though each a_i is calculated twice, it avoids problems caused by large arrays of extended precision numbers. Keep track of the number of times the iteration is performed, so that you know which t gives $(a_{2t} - a_t, N)$ a nontrivial factor of N.

Write a program to implement this for values of N with say eight digits and test the method by multiplying together two or three primes of appropriate size to obtain values for N. You will probably be impressed at the speed of the method, at any rate for numbers around this size. Naturally it is possible to make it more efficient by due cunning; a version of this method was used by R.P.Brent and J.M.Pollard to factorize the huge Fermat number $2^{256} + 1$ in 1980. The two prime factors of this number have 16 and 62 digits!

Why is this called the ρ-method? See Fig. 3.1!

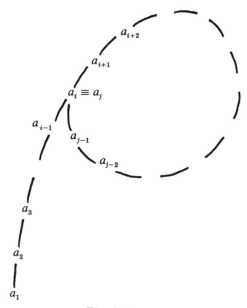

Fig. 3.1

4

Powers and pseudoprimes

1. Fermat's (little) theorem

We begin with a theorem that was discovered by P. de Fermat about 1640; as with many of Fermat's results, no proof by him survives, the first published proof being by L. Euler in 1736. The theorem concerns the powers a, a^2, a^3, ... of a fixed number a modulo a prime p. Since, modulo p, there are only p different values for a^k, there must eventually be a repetition: $a^k \equiv a^j$ for some $k > j$.

1.1 Fermat's theorem *Let p be prime and a be any integer. Then $a^p \equiv a$ (mod p). If $p \nmid a$ then $a^{p-1} \equiv 1$ (mod p).*

Proof Suppose first that $p \nmid a$, so that $(a, p) = 1$, and consider the $p - 1$ numbers

$$a, 2a, 3a, \ldots, (p-1)a.$$

None of these is a multiple of p ($p \mid ja$ implies $p \mid j$ since p is prime and $p \nmid a$). Also no two are congruent modulo p, since $ja \equiv ka$ (mod p) implies $j \equiv k$ (mod p). Thus, modulo p, all these $p - 1$ numbers are different and nonzero. So they must be, modulo p, the numbers $1, 2, 3, \ldots, p - 1$ in some order. Multiplying together we get

$$a^{p-1} 1 \cdot 2 \cdot 3 \ldots (p-1) = a \cdot 2a \cdot 3a \ldots (p-1)a \equiv 1 \cdot 2 \cdot 3 \ldots (p-1)(\bmod\ p).$$

Now the numbers $2, 3, \ldots, p - 1$ are all coprime to p and so can be cancelled one by one from this congruence. It follows that $a^{p-1} \equiv 1$ (mod p). Multiplying by a gives $a^p \equiv a$ (mod p).

If on the other hand $p \mid a$, then a^p and a are both $\equiv 0$ (mod p), so $a^p \equiv a$ (mod p) follows there as well. □

Notice that Fermat's theorem can be construed as saying that the congruence $x^{p-1} \equiv 1$ (mod p) has exactly $p-1$ solutions (namely $x = 1, 2, \ldots, p-1$) which are distinct mod p. This is a special case of a result to be proved below (see 1.4).

1.2 Exercises

(a) Use $2^{12} \equiv 1$ (mod 13) to deduce that $4 \cdot 2^{70} \equiv 1$ (mod 13) and hence that $2^{70} \equiv 10$ (mod 13). Similarly show $3^{70} \equiv 3$ (mod 13). Notice that this shows $2^{70} + 3^{70}$ is divisible by 13.

(b) In a similar way to (a), show that $2^{98} + 3^{98}$ is divisible by 13, and that $2^{100} + 3^{100}$ is divisible by 97.

(c) Starting from $n^7 \equiv n$ (mod 7) deduce that $n^{13} \equiv n$ (mod 7), for any n. Show similarly that $n^{13} \equiv n$ modulo 5, 3 and 2. Deduce that, for every n, $n^{13} - n$ is divisible by $2 \cdot 3 \cdot 5 \cdot 7 \cdot 13 = 2730$.

(d) Starting from $2^{16} \equiv 1$ (mod 17), show that $2^{10^6} \equiv 1$ (mod 17).

(e) Let $f(n) = n^6 + 1091$. Show that if $3 \nmid n$ then $3 \mid f(n)$ and that if $7 \nmid n$ then $7 \mid f(n)$. Deduce that, if n is not a multiple of 42, then $f(n)$ is composite. Show also that, if $n \equiv \pm 2$ (mod 5) then $5 \mid f(n)$, and deduce that the only possible values of n for which $f(n)$ is prime are those of the form $210k$ or $210k \pm 84$. (In fact, according to [Shanks (1971)], only one value of n below 4000, namely 3906, gives a prime value of $f(n)$. You might be able to eliminate some values of k by further cunning, but the identification of the first value $n = 210 \cdot 19 - 84$ giving $f(n)$ prime is rather tricky. See also Chapter 11, 1.8.)

(f) Let $n = r^4 + 1$. Use Fermat's theorem to show that n can *never* be a multiple of 3, 5 or 7. [For example 5: $r^4 \equiv 1$ (mod 5) whenever $5 \nmid r$. So $n \equiv 2$ (mod 5) when $5 \nmid r$ and $n \equiv 1$ (mod 5) when $5 \mid r$.] Can you eliminate any other prime divisors of n? (In fact the smallest prime that can be a factor is 17. The fact that only primes of the form $4m + 1$ can be factors of n follows from Chapter 11, 1.5(a). This can be improved to $8m + 1$.)

1.3 Exercise: Fermat quotients

If p is prime and $p \nmid a$ then, by Fermat's theorem, $(a^{p-1} - 1)/p$ is an integer. It is called the *Fermat quotient* and denoted by $q_p(a)$. Here are some easy properties to verify.

(a) Suppose that $p \nmid a$ and $p \nmid b$. Use $(a^{p-1} - 1)(b^{p-1} - 1) \equiv 0$ (mod p^2) to show that $q_p(ab) \equiv q_p(a) + q_p(b)$ (mod p).

(b) Use the binomial theorem to show that $q_p(p-1) \equiv 1$ (mod p) and $q_p(p+1) \equiv -1$ (mod p).

We conclude this section with a result whose proof could be omitted on first reading; we shall use the result in Chapter 8.

1.4 Theorem   *When p is prime, $d \geq 1$ and $d \mid (p-1)$, the equation $x^d \equiv 1$ (mod p) has exactly d solutions (mod p).*

1.5 Lemma   *Let p be prime and $f(x) = a_0 x^n + a_1 x^{n-1} + \ldots + a_n$ where $p \nmid a_0$. Then the number of solutions of $f(x) \equiv 0$ (mod p) is at most n.*

Note that the hypothesis that p is prime is necessary. For example, $x^2 + x \equiv 0$ (mod 6) has four solutions $x \equiv 0, 2, 3, 5$ (mod 6). What about the hypothesis that $d \mid p - 1$ in 1.4?

Proof of lemma   All congruences will be mod p. When $n = 1$ the result holds because there is actually a unique solution to $a_0 x \equiv -a_1$ when $(a_0, p) = 1$. Assume that the result holds for polynomials of degree $< n$ (where $n \geq 2$) and let f be as in the statement of the lemma. If $f(x) \equiv 0$ has no solution then there is nothing to prove, otherwise let x_1 be a solution. Writing out $f(x) - f(x_1)$ and taking out a factor $x - x_1$ from each term we get

$$(x - x_1)(\text{some polynomial } g(x) \text{ of degree} \leq n - 1) \equiv 0. \qquad (1)$$

The solutions of this are precisely those of $f(x) \equiv 0$ since $f(x_1) \equiv 0$. (Thus a version of the remainder theorem holds here.) *Since p is prime,* (1) is equivalent to $x - x_1 \equiv 0$ or $g(x) \equiv 0$. Since g has degree $< n$, $g(x) \equiv 0$ has $\leq n - 1$ solutions, and so $f(x) \equiv 0$ has $\leq n$ solutions, as required.   □

Proof of 1.4   This is now straightforward, for we write $p - 1 = kd$ so that

$$x^{p-1} - 1 = (x^d - 1)(x^{(k-1)d} + x^{(k-2)d} + \ldots + x^d + 1).$$

The congruence $x^{p-1} - 1 \equiv 0$ has exactly $p - 1 = kd$ solutions, by Fermat's theorem, while the first factor on the right-hand side gives *at most d* solutions and the second factor gives *at most $kd - d$* solutions, by the lemma. Hence the factors on the right must actually give *exactly* d and $kd - d$ solutions, respectively.   □

1.6 Exercises (a) Prove directly the result of 1.4 in the case of an odd prime p and $d = 2$.

(b) Find the solutions of $x^d \equiv 1$ (mod 11) for $d = 2$ and $d = 5$.

(c) Find the solutions of $x^d \equiv 1$ (mod 19) for $d = 2, 3, 6$ and 9.

(d) Prove from 1.4 that, if $d \mid p - 1$, and $a^d \equiv 1$, while $a^k \not\equiv 1$ for $1 < k < d$, then the d different solutions (mod p) of $x^d \equiv 1$ are precisely $x = 1, a, a^2, \ldots, a^{d-1}$.

2. The power algorithm

If we are to make any use of Fermat's theorem, especially of the possibility of proving a number is *not* prime by showing that, for some a with $n \nmid a$, we have $a^{n-1} \not\equiv 1$ (mod n), then we must be able to work out large powers of numbers to a given modulus. The worst possible method would be to work out a^n first and then reduce modulo n: a^n will be huge if n is any size at all. Slightly less crazy would be to work out the successive powers $a, a^2, a^3, \ldots$ and to reduce these modulo n whenever possible. (Notice that we are using here the fact that if $a^k \equiv b$ (mod n) then $a^{k+1} \equiv ab$ (mod n).) But there is another way that is so vastly superior to this that it enables very large powers to be worked out very fast indeed.

As an example, let us evaluate 7^{50} (mod 11). First work out the *binary* expansion of 50:

$$
\begin{array}{rcl}
50 & = & 25 \cdot 2 + 0 \\
25 & = & 12 \cdot 2 + 1 \\
12 & = & 6 \cdot 2 + 0 \\
6 & = & 3 \cdot 2 + 0 \\
3 & = & 1 \cdot 2 + 1 \\
1 & = & 0 \cdot 2 + 1
\end{array}
$$

Thus the binary expansion is $110\,010$. (The entries 0 and 1 are called the 'bits' of the expansion.) Thus

$$= 0 \cdot 1 + 1 \cdot 2 + 0 \cdot 2^2 + 0 \cdot 2^3 + 1 \cdot 2^4 + 1 \cdot 2^5, \text{ so that}$$

$$50 = 7^0 \cdot 7^2 \cdot 7^0 \cdot 7^0 \cdot 7^{16} \cdot 7^{32}.$$

To work out $r \equiv 7^{50}$ (mod 11) we set $a = 7, r = 1$ and square a at every step, so that $a = 7^2, 7^4, 7^8, 7^{16}, \ldots$, multiplying the current r by the current a whenever there is a '1' in the binary expansion of 50. Putting the results in a table gives (writing d for the bit in the expansion):

d	r	$r \pmod{11}$	a	$a \pmod{11}$	
0	1	1	7^2	5	($d = 0$ so r stays at 1)
1	7^2	5	7^4	3	($d = 1$ so $r := r \cdot a = 1 \cdot 5$)
0	7^2	5	7^8	9	($d = 0$ so r unchanged)
0	7^2	5	7^{16}	4	($d = 0$ so r unchanged)
1	7^{18}	9	7^{32}	5	($d = 1$ so $r := r \cdot a = 5 \cdot 4$)
1	7^{50}	1	7^{64}	3	($d = 1$ so $r := r \cdot a = 9 \cdot 5$)

so finally $7^{50} \equiv 1 \pmod{11}$.

Similarly the binary expansion of 24 is 11000 and $7^{24} = 7^8 \cdot 7^{16}$. If we work this out mod 25, then $7^2 \equiv 1$ so all successive powers $7^4, 7^8, \ldots$ are 1 and $7^{24} \equiv 1 \pmod{25}$. Note that this has the form $a^{p-1} \equiv 1 \pmod{p}$ where p is *not* prime!

2.1 Exercise

Show that $3^{50} \equiv 9 \pmod{20}$. Find the binary expansion of 341 and verify that $2^{340} \equiv 1 \pmod{341}$. (The only remotely hard step is $32^2 \equiv 1 \pmod{341}$.) Note that this is another composite number ($341 = 11.31$) satisfying the *conclusion* of Fermat's theorem for $a = 2$.

It is time to formalize the algorithm implicit in the above examples. It is as follows:

2.2 Power algorithm

We seek $r \equiv a^n \pmod{m}$. If n has binary expansion $n = d_k d_{k-1} \ldots d_1 d_0$ where each 'binary digit' or 'bit' d_i is 0 or 1, and $d_k = 1$, then a^n is the product of terms a^{2^i} over those i for which $d_i = 1$. To find this we work out by successive squaring and reduction mod m the numbers $a^2, a^4, a^8, \ldots$. We start with $r = 1$ and multiply the current r by the current power of a whenever the bit d_i is 1. In algorithmic form:

Let $r := 1$;

While $n > 0$, repeat the following:

Let $d := n - 2[n/2]$ (d is the 'units' bit of the current n);

If $d = 1$ then replace $r := ar$ and reduce mod m (otherwise r is unchanged);

Replace $a := a^2$ and reduce mod m;

Replace $n := (n - d)/2$ (the units bit of the new n is the next-to-last bit of the old n);

At the end, when n has become 0, $r \equiv a^n \pmod{m}$.

Turning the algorithm into a PASCAL program gives the following. It is worth, at this stage, writing a procedure which can be used later in

other programs. So we shall call the power algorithm procedure PROCE-
DURE PowerRule, and incorporate this procedure in a program P4_2_3.
(The best course is to name the PROCEDURE PowerRule as a separate
program, say as Power.pas. It can then be used in any other PASCAL
program provided the version of PASCAL being used has a separate
compilation directive, using {$I Power.pas}. But we do not assume this
facility here.)

2.3 Program for the power algorithm

```
PROGRAM P4_2_3;
VAR
a, asaved, n, nsaved, r, m :   extended;
  PROCEDURE PowerRule (base, power, modulus:extended;
    VAR res:extended);
  VAR d:   extended;
  BEGIN
    res:=1;
    WHILE power>0 DO
      BEGIN
        d:=power-2*INT(power/2);
        IF d=1 THEN
          BEGIN
            res:=base*res;
            res:=res-modulus*INT(res/modulus);
          END;
        base:=base*base;
        base:=base-modulus*INT(base/modulus);
        power:=(power-d)/2;
      END;
  END;   (* of PowerRule *)
 BEGIN   (* of main program *)
  writeln;
  writeln ('Type the values of a, n and m,
    separated by spaces');
  readln (asaved, nsaved, m);
  n:=nsaved;
  a:=asaved;
  writeln;
  PowerRule(a,n,m,r);
  writeln;
  writeln ('The value of (', asaved:0:0, ' to the
```

```
        power ', nsaved:0:0,') mod ', m:0:0, 'is ', r:0:0)
    END.    (* of main program *)
```

Note The maximum value of the modulus which you can expect to use safely is about the square root of the maximum precision (so about 10^9 for extended precision). This is because numbers are multiplied together which might in principle be about as big as the modulus. We shall see shortly how to use a different multiplication technique which enhances the precision considerably.

2.4 Computing exercises

(a) Use the program to work out 7^{24} (mod 25) and 2^{340} (mod 341) and check against the answers obtained before. Also check that $5^{280} \equiv 67$ (mod 561), $5^{560} \equiv 1$ (mod 561) and that $100^{560} \equiv 1$ (mod 561).

(b) Find 17^{3313} and 3313^{17} (mod 112 643). What can you deduce about $\gcd(p^q - 1, q^p - 1)$, where $p = 17$ and $q = 3313$? (Be just a little careful in your answer. Since $p \equiv 1$ and $q \equiv 1$ (mod 16) the gcd is certainly divisible by 16.) (Compare Chapter 6,2.9 below.) What can you say about $\gcd((p^q - 1)/(p - 1), (q^p - 1)/(q - 1))$?

(c) Verify (if precision permits) that $3^{1\,373\,652} \equiv 1$ (mod 1 373 653). Also find the factorization of 1 373 653 to check that it is *not* prime. Check that $a^m \equiv a$ (mod m) and that m is not prime, in the following cases:

$$(a, m) = (2, 161\,038), (2, 215\,326), (2, 314\,821), (3, 314\,821),$$

$$(5, 314\,821), (7, 314\,821).$$

In which of these is it also true that $a^{m-1} \equiv 1$ (mod m)?

(d) Find the solutions of $x^4 \equiv 1$ (mod 821). (Since 821 is prime, there will be exactly four of them, by 1.4, and two are easy to find.) You might want to add a loop to the program above, to save some tedium.

(e) Let $m = 37 \cdot 2^{16} + 1 = 2\,424\,833$. Check that $37^{32} \equiv -1$ (mod m). Now use the congruence $37 \cdot 2^{16} \equiv -1$ (mod m), raised to the power 32, to deduce that m is a factor of the Fermat number $F_9 = 2^{2^9} + 1$. In fact m is prime, as you can verify with P2_1_13. (This factor was found by A.E. Western in 1903, but the complete factorization of F_9 was found only in 1990; see [Cipra (1990)].)

(f) In a similar manner to (e), let $m = 11\,131 \cdot 2^{12} + 1 = 45\,592\,577$. By raising $11\,131 \cdot 2^{12} \equiv -1$ (mod m) to a suitable power, show that m is a factor (also prime) of the Fermat number $F_{10} = 2^{2^{10}} + 1$. Note that this Fermat number has 309 digits! This factor was found by J.L. Selfridge in 1953, but the full factorization of F_{10} is not known.

(g) There are many other examples of factors $k.2^n + 1$ of Fermat numbers in [Riesel (1985), pp.377–9], some of which are accessible to direct attack such as in the previous two exercises. For example, can you find a small number k such that $m = k \cdot 2^{25} + 1$ is a factor of F_{23}? (So you will want to raise $k \cdot 2^{25} \equiv -1 \pmod{m}$ to a suitable power so that the exponent 25 becomes as near as possible to $2^{23} = 8\,388\,608$.) You can verify that the resulting factor m is prime using P2_1_13 or P2_1_15.

2.5 Project: Pollard's $p-1$ method of factoring One version of this factoring method is as follows. Suppose that p is prime and k is such that $(p-1) \mid k!$. Why does this imply that all the prime factors of $p-1$ are $\leq k$? Why does Fermat's theorem imply that $2^{k!} \equiv 1 \pmod{p}$?

Suppose we wish to factor a given integer N. Take a small integer, say 2, and compute successively

$$2^1, 2^2, 2^6, 2^{24}, \ldots, 2^{k!} \text{ all } (\bmod N),$$

where k is reasonably large (but not too large!). If p is a prime factor of N then certainly p is a factor of $(\langle 2^{k!} - 1 \rangle_N, N)$ and, with luck, this gcd will be p rather than N. (Recall the notation $\langle x \rangle_N$ for the least positive residue of $x \bmod N$.) Write a program to implement this method and test it on numbers which are the product of two primes. Note that for success we expect to require all the prime factors of the sought factor p to be $\leq k$.

3. Head's algorithm for multiplication mod m

It is not necessary to work through all the theory of this section but to handle the largest numbers in the chapters which follow it is necessary that you use the PROCEDURE in 3.4 below which implements the algorithm. If we multiply two numbers modulo m in the ordinary way, then clearly we need to handle numbers almost as large as m^2. However multiplication is just repeated addition so in principle it must be possible to effect the multiplication without handling numbers larger than $2m$. (For example, 3×8 modulo 10 could be worked out as $8 + 8 = 16 \equiv 6, 6 + 8 = 14 \equiv 4$.) Of course this would be incredibly impractical for numbers of ten digits, and Head's algorithm (published in 1980) is a more subtle way of doing the multiplication without introducing numbers bigger than $2m$. In the form we shall implement it, $2m$ here is replaced by $4m$, which is hardly to be worried about. With extended precision, this enables numbers as large as 10^{18} or so to be handled with confidence. Head's algorithm can be incorporated as a procedure into the power algorithm, and that is what we recommend you do, even if you do not take the trouble to work through the proof of the multiplication algorithm which follows.

We seek to multiply x and y modulo m, where we assume $0 \le x < m$, $0 \le y < m$. The method proceeds in several stages, and we shall leave some of the details of checking as exercises.

3.1 Lemma *Let $m \ge 2$, $T = [\sqrt{m}+\frac{1}{2}]$, $t = T^2 - m$. Then $-T \le t \le T$ and $T^2 \le 2m$.*

Proof From the definition of T, we have

$$\sqrt{m} - \tfrac{1}{2} < T \le \sqrt{m} + \tfrac{1}{2} \tag{1}$$

and squaring these inequalities gives

$$m - \sqrt{m} + \tfrac{1}{4} < T^2 \le m + \sqrt{m} + \tfrac{1}{4}. \tag{2}$$

Adding the two left-hand inequalities of (1) and (2) gives $m - \frac{1}{4} < T^2 + T$ and since the right-hand side is an integer, we get $T^2 + T \ge m$, which is part of what is required.

The second part comes from adding the right-hand inequality of (2), and the left-hand inequality of (1) multiplied by -1: we get $T^2 - T < m + \frac{3}{4}$, which gives $T^2 - T \le m$. Finally, $T^2 \le 2m$ follows easily from the right-hand half of (2). $\qquad\square$

3.2 Lemma *Let $x = aT + b$, where $0 \le b < T$. Then $0 \le a \le T$. Let $y = cT + d$, where $0 \le d < T$. Then $0 \le c \le T$.*

Proof Clearly $a \ge 0$. If $a > T$ then $x = aT + b \ge aT \ge (T+1)T \ge m$, by 3.1. This is a contradiction. The other half is similar.

3.3 Lemma (a) *Define $z \equiv ad + bc \pmod{m}$, where $0 \le z < m$. Then $ad \le m$, $bc \le m$ and*

$$xy \equiv act + zT + bd \pmod{m}.$$

(b) *Let $ac = eT + f$, where $0 \le f < T$. Then $0 \le e \le T$ and*

$$xy \equiv (et + z)T + ft + bd \pmod{m}.$$

(c) *Let $v \equiv z + et \pmod{m}$, $0 \le v < m$.*
Let $v = gT + h$, where $0 \le h < T$. Then $0 \le g \le T$ and

$$xy \equiv hT + (f + g)t + bd \pmod{m}.$$

Proof All these are easy to check, or follow similarly to the results in 3.2. For instance, if $e > T$ in (b) then $ac \ge eT > T^2$, which is false by 3.2. $\qquad\square$

All the steps in working out in succession (all $\equiv$ are mod m)

$$a, b, c, d; z, e, f; v; g, h; j \equiv (f + g)t, k \equiv j + bd; xy \equiv hT + k$$

involve numbers of size at most $4m$. Here is a PASCAL procedure, called
MultiplyModM, for effecting this algorithm. It can be incorporated into
the program P4_2_3 for evaluating powers, replacing the explicit mul-
tiplications by calls to the procedure. We give the details below, as
program P4_3_4. Note that since t can be < 0 we have to make al-
lowance for the irritating fact that INT in PASCAL rounds 'towards
zero' instead of 'to the left'.

3.4 PROCEDURE for implementing Head's algorithm

```
PROCEDURE MultiplyModM (x,y,modulus :  extended;
VAR prod :  extended);
(* Note that in most examples it will be the same *)
(* modulus m throughout a calculation, so it is more *)
(* efficient to calculate T (here called capT) and *)
(* t (see 3.1) once during the main program than *)
(* to calculate them within the procedure each time *)
(* they are used.  The result, prod, is xy mod m *)
VAR
a, b, c, d, z, e, f, v, v1, g, h, j, j1, k:  extended;
(* v1 and j1 are extra variables used in the *)
(* calculation when allowing for t to be <0 *)
BEGIN
  a:= INT(x/capT); b:= x - a * capT;
  c:= INT(y/capT); d:= y - c * capT;
  z:= a * d + b * c; z:= z - modulus * INT(z/modulus);
  e:= INT(a*c/capT); f:= a * c - e * capT;
  v:= z + e * t;
  v1:= v/modulus;
  IF (v>0) OR (v1=INT(v1)) THEN v:= v - modulus * INT(v1)
    ELSE v:=v - modulus * (INT(v1) - 1);
    (* This allows for v < 0 *)
    g:= INT(v/capT); h:= v - g * capT;
    j:= (f + g) * t;
    j1:=j/modulus;
    IF (j>0) OR (j1=INT(j1)) THEN j:= j - modulus
    * INT(j1)
    ELSE j:= j - modulus * (INT(j1) - 1);
  k:= j + b * d; k:= k - modulus * INT(k/modulus);
  prod:= h * capT + k; prod:= prod - modulus
    * INT(prod/modulus);
END;    (* of the PROCEDURE MultiplyModM *)
```

Incorporating the above procedure into program P4_2_3 gives the following.

```
PROGRAM P4_3_4;    (* PowerRule using Head's algorithm *)
VAR
a, asaved, n, nsaved, r, m, capT, t:extended;

  PROCEDURE MultiplyModM ({inputs} x,y,modulus:
          extended; {output}VAR prod:  extended);
```

Copy here the procedure given above

```
  END;    (* of MultiplyModM *)

  PROCEDURE PowerRule (base, power, modulus:extended;
        VAR res:  extended);
  VAR d:  extended;

  BEGIN
    res:=1;
    WHILE power>0 DO
      BEGIN
        d:=power-2*INT(power/2);
        IF d=1 THEN
          MultiplyModM(base,res,modulus,res);
          (* Here we use Head's algorithm *)
        MultiplyModM(base,base,modulus,base);
        (* also here *)
        power:=(power-d)/2;
      END;

  END;    (* of PowerRule *)


BEGIN    (* of main program *)
  writeln;
  writeln ('Type the values of a, n and m separated
    by spaces');
  readln (asaved, nsaved, m);

  capT:=INT(SQRT(m)+0.5);
  t:=capT*capT-m;
  n:=nsaved;
  a:=asaved;

  writeln;
```

```
PowerRule(a,n,m,r);

writeln;
writeln ('The value of (', asaved:0:0, ' to the
power ', nsaved:0:0,') mod ', m:0:0, ' is ', r:0:0);

END.   (* of main program *)
```

You should of course test the resulting program with the examples given above for the ordinary power algorithm. A good test with larger numbers is to take $a = 2$ and $m =$ any large prime, $n = m - 1$. The residue, by Fermat's theorem, should be 1. Here are a handful of largish primes: $10^9 + 7, 10^{10} + 19, 10^{11} + 3, 10^{12} + 39, 10^{13} + 37, 10^{14} + 31, 10^{15} + 37$.

3.5 Computing exercises   (a) Let $m = 2^{2^5} + 1$, $a = 3$, $n = m - 1$. Verify that $a^n \not\equiv 1 \pmod{m}$. This shows, of course, that m, the fifth Fermat number, is not prime. Later on we shall give Pepin's test for the primality of $F_k = 2^{2^k} + 1$; this exercise is an example of that test.

(b) It is rather rare for a^{p-1} to be congruent to 1, not merely mod p as Fermat's theorem requires, but mod p^2. Primes for which this holds with $a = 2$ are called 'Wieferich primes' in [Ribenboim (1988), Sec. 5.III], since in 1909 A. Wieferich considered such primes in connexion with 'Fermat's last theorem'. A great deal is not known about such primes, such as whether there are infinitely many of them. Armed with the power algorithm in the above form, you can verify the following for yourself.

The congruence $a^{p-1} \equiv 1 \pmod{p^2}$ holds for the following pairs (a, p) (for the biggest ones you *do* need extended precision!):

$$(2, 1093), (2, 3511);$$

$$(19, 3), (19, 7), (19, 13), (19, 43), (19, 137), (19, 63\,061\,489);$$

$$(41, 29), (41, 1\,025\,273), (41, 138\,200\,401).$$

Of course, there is nothing to stop you verifying that those big numbers *are* primes, using the techniques of Chapter 2.

Note that, for a *given* p, the solutions of $a^{p-1} \equiv 1 \pmod{p^2}$ can be found by the method of 'primitive roots'. See Chapter 8, 1.2 below.

3.6 Project: Wieferich primes and more   By adding an appropriate loop to the power algorithm program, verify that, for $a = 2$, the only primes $p < 4000$ which satisfy $a^{p-1} \equiv 1 \pmod{p^2}$ are those in 3.5(b) above. (In fact no other solution with $a = 2$ is known.) Likewise for $a = 19$ show that the only solutions with say $p < 1000$ are those given above. For $a = 53$, find the four primes < 100 which satisfy the congruence.

4. Pseudoprimes

We have had several examples of numbers which 'masquerade as primes' in the sense that they are composite but satisfy the *conclusion* of Fermat's theorem. (See 2.1,2.4(a),2.4(c).)

4.1 Definition   *Let b be a positive integer. A positive integer n is said to be a* pseudoprime to the base b *provided n is composite, and* $b^n \equiv b$ (mod n). Note that, if $(b, n) = 1$, then n is a pseudoprime to the base b if and only if n is composite and $b^{n-1} \equiv 1$ (mod n). In fact, some authors include $(b, n) = 1$ in the definition of pseudoprime to the base b, but we shall use the more general definition.

Clearly, pseudoprimality to base 1 is not interesting.

4.2 Computing exercises   (a) Verify that 341, 561 and 645 are pseudoprimes to the base 2. (Don't forget to check that they are composite!)

(b) Verify that 161 038 and 215 326 are pseudoprimes to the base 2. Note that they are *even*. In fact they are the smallest even pseudoprimes to the base 2.

(c) Verify that 1105 is a pseudoprime to bases 2 and 3, that 1729 is a pseudoprime to bases 2, 3 and 5, and that 29 341 is a pseudoprime to bases 2, 3, 5 and 7.

(d) Determine the first number $> 10^{15}$ (or 10^{10} if you work without extended precision) which is *either* prime *or* a pseudoprime to base 2. Do you get the same answer for base 3? (If so, then it is quite likely that you have hit on a prime.) If you were just looking for the first 'probable prime' $> 10^{15}$, in the sense of the first number which is a prime or a pseudoprime and not obviously composite, then of course you would omit multiples of 2, 3 and 5 and so shorten the calculations. We shall meet many calculations along these lines in the next chapter.

4.3 Exercises

(a) Show that, if $(a, n) = 1$ and n is a pseudoprime to bases a and ab, then n is a pseudoprime to base b as well.

(b) Let p be prime. Use Fermat's theorem to show that $2^p - kp - 2 = 0$ for some integer k. Writing $r = 2^p - 2$, show that $2^r - 1 = (2^p - 1)(2^{r-p} + 2^{r-2p} + \ldots + 1)$ and deduce that $2^p - 1$ is either a prime or a pseudoprime to base 2. (Numbers $2^p - 1$ are called Mersenne numbers. Compare Chapter 1, 1.5(c).)

4.4 Project: pseudoprimes to more than one base   Write a program to show that the numbers given in 4.2(c) above are the *smallest* pseudoprimes to the given multiple bases. (Don't forget that primes are *not* allowed.)

4.5 Project: pseudoprimes to the base 2 Write a program to find all the pseudoprimes to base 2 which are < 2000. (If it weren't for weeding out the true primes, this would be a very easy project.)

4.6 Theorem *There are infinitely many pseudoprimes to any given base.*

Proof Let a be an integer > 1 and p be an odd prime with $p \nmid a$, $p \nmid (a^2 - 1)$. Clearly there are infinitely many such primes p , for any given a. Let

$$n = \frac{a^{2p} - 1}{a^2 - 1} = a^{2p-2} + a^{2p-4} + \ldots + a^2 + 1.$$

We claim that n is a pseudoprime to base a; this will prove the result. First, $n-1$ is a sum of $p-1$ terms, that is an even number of terms since p is odd, and each of those terms has the parity (evenness or oddness) of a. Hence $n-1$ is even.

Next, rearranging the above fraction gives $(n-1)(a^2-1) = a^2(a^{2p-2} - 1)$. Now $a^{2p-2} = (a^{p-1})^2 \equiv 1 \pmod{p}$ by Fermat's theorem, and $p \nmid (a^2 - 1)$ by hypothesis, so $p \mid (n - 1)$. Since p is odd and $n - 1$ is even, $2p \mid (n - 1)$, $n - 1 = 2pk$ say.

Finally, cross-multiplying the above fraction gives $a^{2p} \equiv 1 \pmod{n}$, so that $a^{n-1} = (a^{2p})^k \equiv 1 \pmod{n}$, as required. (Note that $(a, n) = 1$ *follows* from this, or from the expression for n as a sum.)

Well, not quite finally, for we have to prove that n is composite! In fact, $n = \frac{a^p - 1}{a-1} \times \frac{a^p + 1}{a+1}$. Both factors are integers (and > 1) since p is odd. □

4.7 Exercises

(a) Take $a = 3$ and the smallest allowable value for p, and hence find a pseudoprime to the base 3.

(b) Here is the outline of a different proof that there are infinitely many pseudoprimes to the base 2. In fact it shows that if n is an odd pseudoprime to base 2, then so is $N = 2^n - 1$. Note that, since such an n is composite, then so is N, using the first factorization of Chapter 1, 1.5(c). Use Fermat's theorem, as in 4.3(b) above, to show that $2^n - 2 = kn$ for some k, and deduce that $N \mid (2^{N-1} - 1)$. (Use again the same factorization from Chapter 1.)

(c) Could $2^n - 2$ ever be a pseudoprime to base 2? This question was answered affirmatively by W.L. McDaniel [McDaniel (1989)]. Let $n = 465\,794$; verify using the power algorithm that $2^n \equiv 3 \pmod{n - 1}$.

Deduce that

$$(2^{n-1} - 1) \mid (2^{2^n - 3} - 1)$$

and hence that, with $N = 2^n - 2$, we have $2^N \equiv 2 \pmod{N}$, so that indeed N is a pseudoprime to base 2.

Despite (b) above, pseudoprimes to base 2 are much rarer than primes. In fact there are $882\,206\,716$ primes less than $2 \cdot 10^{10}$ but only $19\,865$ pseudoprimes to base 2 in this range. So if a number appears to be a pseudoprime to base 2 then there is a 'very good chance' that it is prime! Of course one can increase the likelihood by testing for pseudoprimalty to more than one base, such as 2, 3, and 5. (Compare 4.2(d) above.) There are numbers which will slip through that net too (see 4.2(c)). In fact there are numbers which will slip through all the pseudoprime tests and yet are composite, as we now show.

4.8 Definition   *A composite number n which satisfies $b^{n-1} \equiv 1 \pmod{n}$ for all b which are coprime to n is called a* Carmichael number *(after R.D. Carmichael, who introduced them in 1912).*

Of course making a definition does not prove that such numbers exist, but here is an example. Let $n = 561 = 3 \cdot 11 \cdot 17$. Thus $(b, 561) = 1$ implies that $(b, 3) = (b, 11) = (b, 17) = 1$. Using Fermat's theorem,

$b^2 \equiv 1 \pmod 3$ implies $b^{560} = (b^2)^{280} \equiv 1 \pmod 3$;
$b^{10} \equiv 1 \pmod{11}$ implies $b^{560} = (b^{10})^{56} \equiv 1 \pmod{11}$;
$b^{16} \equiv 1 \pmod{17}$ implies $b^{560} = (b^{16})^{35} \equiv 1 \pmod{17}$.

Since 3, 11 and 17 are three distinct primes, this implies $b^{560} \equiv 1 \pmod{561}$.

We leave it as an exercise to apply exactly the same method to the general result:

4.9 Proposition   *Let $n = q_1 q_2 \ldots q_k$, where the q_i are distinct primes and $k \geq 2$. Suppose that, for each $i, (q_i - 1) \mid (n - 1)$. Then n is a Carmichael number. (Note that the primes must be odd since if $q_1 = 2$ then n is even, and, q_2 being odd, it is not possible for $(q_2 - 1) \mid (n - 1)$.)*

As it happens, all Carmichael numbers are generated in this way; see Chapter 8,2.14.

4.10 Exercises

(a) Suppose that p, $2p-1$ and $3p-2$ are all primes, where $p > 3$. Show that their product is a Carmichael number. You will need to show that $p \equiv 1 \pmod 6$; remember that any prime > 3 is $\equiv \pm 1 \pmod 6$. In view of this the three primes can be written in the more pleasing form $6m + 1$,

$12m+1$, $18m+1$, and $m=1$ gives the first example $7 \cdot 13 \cdot 19$. Find the next value of m which gives three primes $6m+1$, $12m+1$, $18m+1$.

(b) Suppose that $k=2$ in 4.9 and let $q_2 > q_1$. Show that $n-1 \equiv q_1 - 1$ (mod $q_2 - 1$), and derive a contradiction from this. Hence the minimum possible value of k in 4.9 is 3 (as the example before 4.9 shows).

(c) Let n be as in 4.9 (hence a Carmichael number). Show that $b^n \equiv b$ (mod n) for any base b, whether coprime to n or not. [Hint. Assume that $(n,b) > 1$. Why must (n,b) be a product of primes from among the q_i? Show that $b^n \equiv b$ (mod q_i) for each prime q_i occurring in (n,b) and $b^{n-1} \equiv 1$ (mod q_i) for each prime q_i not occurring in (n,b).]

We might hope that results such as (a) above would help us to prove that there are infinitely many Carmichael numbers, but in fact it is usually very hard indeed to decide whether there are infinitely many primes in two or more special sequences. For instance, it is not even known whether there are infinitely many 'prime pairs' $p, p+2$. The question as to whether there are infinitely many Carmichael numbers was settled affirmatively in 1992; see [Granville (1992)].

4.11 Computing exercises
(a) Factorize $29\,341, 172\,081, 564\,651\,361$ and hence verify that they are all Carmichael numbers.

(b) Find a Carmichael number $7 \cdot 13 \cdot 31 \cdot q$, where q is prime, and three Carmichael numbers $13 \cdot 37 \cdot q$, where q is prime. (Of course you should use the criterion 4.9. Probably you will want to access the file created in 1.10 of Chapter 2.)

(c) Verify that $m = 35$ and $m = 1515$ give Carmichael numbers in the manner of 4.10(a) above. (A rather harder exercise is to verify that these are the next smallest values of m after $m = 1$ and the one you (might have) found in 4.10(a).)

(d) Let $p = 187\,687$, $q = 375\,373$, $r = 43\,355\,467$. In fact these are all primes. It is not feasible, even in extended precision, to verify that $n = pqr$ is a Carmichael number since pqr is too big. However, show that the conditions that $n-1$ is divisible by $p-1$, $q-1$ and $r-1$ are equivalent to:

$$(p-1) \mid (qr-1), \ (q-1) \mid (pr-1), \ (r-1) \mid (pq-1).$$

It is feasible to verify these within extended precision; try doing this. Also replace r by $6\,404\,784\,751$ (also a prime) and do the same thing. These triples have the form $6m+1, 12m+1, 18mk+1$ for suitable values of m and k; compare 4.10(a) above. See also [Dubner (1989)].

4.12 Project: Carmichael numbers Find all 16 Carmichael num-

bers which are < 10000. (Don't forget the result of 4.10(b) above.) Also find all the Carmichael numbers which are products of three primes, all of which are < 100.

5. Wilson's theorem

We conclude this chapter with a theorem which is spectacularly useless in the search for primes, but which nevertheless has a certain charm. The 'only if' part of the theorem is in fact a special case of Chapter 3,3.4, but we shall give the proof here. The theorem was stated by J. Wilson (1741–1793) but the first published proof is by Lagrange in 1770.

5.1 Wilson's theorem   *An integer $p > 1$ is prime if and only if $(p - 1)! \equiv -1 \pmod{p}$.*

Proof Suppose that p is a prime > 2. We have to multiply together all the numbers $1, 2, \ldots, p-1$ and take the residue modulo p. Each of these numbers x has an inverse $y \pmod{p}$, that is $xy \equiv 1 \pmod{p}$. Further we can always reduce $y \pmod{p}$ so that y is also one of the numbers $1, 2, \ldots, p - 1$. We have $x = y \Leftrightarrow x^2 \equiv 1 \pmod{p} \Leftrightarrow x \equiv \pm 1 \pmod{p}$ (since p is prime; compare 1.4(g) in Chapter 3) $\Leftrightarrow x = 1$ or $x = p - 1$. The other numbers pair off with the product of each pair equal to 1 $\pmod{p}$, so $1 \cdot 2 \cdot 3 \ldots (p-1) \equiv 1(p-1) \equiv -1 \pmod{p}$. The result holds for $p = 2$ too!

Suppose on the other hand that n is composite, say $n = ab$ where $a > 1$ and $b > 1$, and $(n - 1)! \equiv -1 \pmod{n}$. We seek a contradiction. Clearly $a \mid (n - 1)!$ (after all $1 < a < n$), and we know $a \mid n$, so the congruence implies $a \mid (-1)$, which is false. (Why does this contradiction disappear if we start out with $n = $ prime?) □

Although Wilson's theorem provides in principle a test for primality it is of course vastly inefficient since it is necessary to multiply all of 2, $3, \ldots, p-1$ together mod p in order to use it. This is slower than testing p for primality by trial division by primes $\leq \sqrt{p}$.

5.2 Exercises

(a) Let $p = 4k + 1$ be prime. Wilson's theorem says that $(4k)! \equiv -1$ $\pmod{p}$. Consider the terms $2k + 1, \ldots, 4k$ which occur in the product giving $(4k)!$ and show that in fact $((2k)!)^2 \equiv (4k)! \pmod{p}$. Deduce that $x^2 \equiv -1 \pmod{p}$ has solutions $x = \pm(2k)!$. (By Lemma 1.5 above, these are the only two solutions.)

(b)* Show that p and $p+2$ are both prime ('twin primes') if and only if $4((p-1)!+1)+p \equiv 0 \pmod{p(p+2)}$. [Hints for 'if'. Show p is odd by assuming p even and using the congruence to obtain a contradiction. Use the congruence to show $(p-1)!+1 \equiv 0 \pmod p$ and deduce p is prime by Wilson's theorem. Now use $4(p-1)!+2 \equiv 0 \pmod{p+2}$ to show that $(p+1)!+1 \equiv 0 \pmod{p+2}$ and deduce that $p+2$ is prime.] A criterion for p and $p+d$ to be both prime, again using Wilson's theorem, is given, with proof, in *College Math. J.* **19** (1988), p. 191.

5.3 Computing Exercise Verify that, for $p = 563$, $(p-1)! \equiv -1$ $\pmod{p^2}$.

5.4 Project: Wilson primes We could call a prime a 'Wilson prime' if $(p-1)!+1$ is congruent to 0 not merely modulo p but modulo p^2. Write a program to determine Wilson primes and in particular verify that the only ones < 1000 are $5, 13$ and 563. (These are in fact the only Wilson primes known. See [Ribenboim (1988), Sec. 5.IV].)

5

Miller's test and strong pseudoprimes

1. Miller's test

Consider the following calculations, all of which start with a verification that some number n is a prime or a pseudoprime to a base b. (You can of course check these using Program 4_2_3.)

$$n = 25, \ b = 7 : 7^{24} \equiv 1, \ 7^{12} \equiv 1, \ 7^6 \equiv -1 \ (\text{mod } 25). \qquad (i)$$

$$n = 2047, \ b = 2 : 2^{2046} \equiv 1, \ 2^{1023} \equiv 1 \ (\text{mod } 2047). \qquad (ii)$$

$$n = 341, \ b = 2 : 2^{340} \equiv 1, \ 2^{170} \equiv 1, \ 2^{85} \equiv 32(\text{mod } 341). \qquad (iii)$$

$$n = 561, \ b = 2 : 2^{560} \equiv 1, \ 2^{280} \equiv 1, \ 2^{140} \equiv 67(\text{mod } 561). \qquad (iv)$$

$$n = 2243, b = 2 : 2^{2242} \equiv 1, 2^{1121} \equiv -1(\text{mod } 2243). \qquad (v)$$

In each case, we keep halving the exponent until one of two things happens: either the exponent is odd (as in (ii), (iii) and (v)), or the residue is different from 1 (as in (i), (iii), (iv) and (v)). In (iii) and (v), both things happen at the same step.

If we start with an *odd prime* for n, what would happen? Certainly $b^{n-1} \equiv 1 \ (\text{mod } n)$ is assured by Fermat's theorem, and $n - 1$ is even, so we can proceed to work out $b^{(n-1)/2} = x$, say. Since $x^2 \equiv 1 \ (\text{mod } n)$ and n is *prime*, it follows that $x \equiv \pm 1 \ (\text{mod } n)$ (compare Chapter 3, 1.4(f)). If we get $+1$ and if $(n - 1)/2$ is still even then we can proceed to work out $b^{(n-1)/4} = y$, say, and since $y^2 \equiv 1 \ (\text{mod } n)$ we have again that $y \equiv \pm 1 \ (\text{mod } n)$. Thus, continuing in this way, we shall end up with either a residue of -1, or an odd exponent (or both at the same step), and then we shall stop. (If $x^2 \equiv -1 \pmod{n}$, where n is prime, there is no very useful prediction of what x will be.)

Note that, for a given prime value of n, the number of steps required to reach a residue of -1 may be very different for different values of b. For example, with $n = 257$, we find that $2^{(n-1)/r} \equiv 1 \pmod{n}$ for $r = 1$, 2, 4, 8, 16, 32, while the residue is -1 for $r = 64$. On the other hand, for the same n, $3^{(n-1)/r} \equiv 1 \pmod{n}$ for $r = 1$ but the residue is -1 for $r = 2$. Of course this means that, for $b = 2$, *a much lower power* of 2 than the $(n-1)^{\text{st}}$ power is congruent to 1 mod n. We shall have much to say about this phenomenon in the next chapter when we deal with *orders modulo* n. For powers of 3, on the other hand, there is in fact *no lower power* than the $(n-1)^{\text{st}}$ which is congruent to 1 mod n. We shall call 3 a *primitive root modulo* n; see Chapter 8. Once you have program P5_1_5 available, you can easily test for yourself the number of steps which any prime takes to reach a residue of -1 by the above process. See 1.8 below.

It is clear that it should be harder for a composite number to masquerade as a prime through all the steps than it is for a composite number merely to satisfy the conclusion of Fermat's theorem (the first step). For example, 341 and 561 are both composite, and they reveal their compositeness in the calculations (iii) and (iv) above. On the other hand, 25 and 2047 ($= 23 \times 89$) are both composite too, and they masquerade successfully as primes in the calculations (i) and (ii) above.

The sequence of calculations sketched above is called *Miller's test*. (The version sketched in section 2 below was published by G.L. Miller in 1976.) A composite number which gets through the test in the same way that a prime number would is called a *strong pseudoprime to base* b. We shall now make these matters more precise. All congruences are modulo n and we use $\langle x \rangle$ to denote the least positive residue of x (mod n).

1.1 Miller's test to base b   Let n be an odd positive integer > 1 and let b be coprime to n.

Step 1: Let $k = n - 1$, $\langle b^k \rangle = r$. If $r = 1$ then continue, otherwise n *fails* the test.

While k is even and $r = 1$, repeat the following:

Step 2: Replace k by $k/2$, and replace r by the new value of $\langle b^k \rangle$.

When k fails to be even or r fails to be 1:

If $r = 1$ or $n - 1$ then n *passes* the test.

If $r \neq 1$ and $r \neq n - 1$ then n *fails* the test.

1.2 Definition   *If n is composite and n passes Miller's test to base b, then n is called a* strong pseudoprime *to base b.*

We hope it is clear from the above discussion that, if n is prime and $n \nmid b$ (so that $(b, n) = 1$), then n will always pass Miller's test to base b.

The crucial fact is that $x^2 \equiv 1 \pmod{n}$ implies $x \equiv \pm 1 \pmod{n}$ when n is prime, so that no residue other than ± 1 (that is, 1 or $n - 1$) can ever arise and Miller's test cannot fail. Of course Step 1 is just Fermat's theorem. Let us record this fact in the following way:

1.3 Proposition   *If n fails Miller's test to base b, then n is composite.*

Notice that the calculations at the beginning of this section show that 25 is a strong pseudoprime to base 7 and that 2047 is a strong pseudoprime to base 2.

1.4 Exercises

(a) Let n be odd, $n \geq 3$. Why will n always pass Miller's test to base $n - 1$? [In congruences modulo n, $n - 1$ can be replaced by -1.]

(b) Suppose that, in performing Miller's test on n to base b, we reach, for some $r \geq 0$, $b^{(n-1)/2^r} \equiv 1 \pmod{n}$, but $c = b^{(n-1)/2^{r+1}} \not\equiv \pm 1 \pmod{n}$, where $2^{r+1} \mid n - 1$. Why has n failed Miller's test to base b? Show that $(c - 1, n)$ is a proper factor of n. [It is clearly a factor; you have to show it is not 1 and not n. Compare Chapter 3,1.5.] In example (iii) at the beginning of this section, $c = 32$, $n = 341$ gives $(c - 1, n) = 31$, while in example (iv), $(c - 1, n) = 33$. This method enables us to find a proper factor of any *pseudoprime* n to base b (where $(b, n) = 1$) which fails Miller's test to base b. (Of course, even if $b^{n-1} \equiv c \not\equiv 1 \pmod{n}$ then it is *conceivable* that $(c - 1, n)$ is a proper factor of n. There are some examples in 1.7(b)–(d) below.)

1.5 Program for Miller's test   Miller's test is very easy to implement using the PowerRule procedure which can be copied directly from Program P4_2_3. The following program performs Miller's test on an odd n to base b.

```
PROGRAM P5_1_5;

VAR n, b, n1, r :  extended;
(* n1 is used to store the current power (called k *)
(* in the discussion) and r is the current residue *)

PROCEDURE PowerRule (base, power, modulus :  extended;
          VAR res :  extended);
```
Copy here the PowerRule procedure from P4_2_3

```
BEGIN (* main program *)
  writeln('Type the values of n and b');
  readln(n,b);
  n1:= n - 1;
  r:= 1;
  WHILE (n1 = INT(n1)) AND (r = 1) DO
    BEGIN
(* If you want to write the current power n1 *)
(* then do it here *)
      PowerRule(b, n1, n, r);
      n1:= n1/2;
      writeln('Residue is', r:0:0);
    END;
  IF (r = 1) OR (r = n - 1) THEN
    writeln(n:0:0, 'has passed Millers test to base',
      b:0:0)
    (* Note no semicolon!Also no ' in Miller's *)
    (* because that's not allowed !*)
  ELSE
    writeln(n:0:0, 'has failed Millers test to base',
      b:0:0);
END.
```

1.6 Program for Miller's test using Head's algorithm   In order to handle the largest integers encountered later, it is necessary to use the version of the power rule which incorporates Head's algorithm, called MultiplyModM in Chapter 4,3.4. This takes the following form.

```
PROGRAM P5_1_6;

VAR n, b, n1, r, capT, t: extended;   (* The extra *)
(* variables capT and t are used in Head's algorithm *)

  PROCEDURE MultiplyModM (x,y,modulus: extended;
        VAR prod: extended);
```
Copy the procedure from Chapter 4,3.4

```
  PROCEDURE PowerRule (base, power, modulus: extended;
        VAR res: extended);
```
Copy the procedure from Program P4_2_3

```
BEGIN   (* main program *)
  writeln('Type the values of n and b');
  readln(n,b);
```

```
capT:= INT(SQRT(n) + 0.5);
t:= capT * capT - n;
```
Now copy the rest of the main program in P5_1_5 above

Note   In what follows, we shall assume that you have available the program P5_1_6 incorporating Head's algorithm. If you do not, then calculations with moduli more than the square root of the precision (so more than about 10^9 with extended precision) will go awry.

1.7 Computing exercises

(a) Check that $n = 1\,373\,653$ passes Miller's test to bases 2 and 3. Now check that n is composite, using the program of Chapter 2,1.13. Hence n is a strong pseudoprime to bases 2 and 3. (It is the smallest such.) Check that n fails Miller's test to base 5, indeed is not even a pseudoprime to base 5: it fails Miller's test at Step 1. Nevertheless check that if $5^{n-1} \equiv c \pmod{n}$ then $(c - 1, n) = 829$, a proper factor of n. (Compare 1.4(b) above.)

(b) Check that $n = 25\,326\,001$ is a strong pseudoprime to bases 2, 3 and 5. (It is the smallest such.) Check that n is not even a pseudoprime to base 7, but that nevertheless $7^{n-1} \equiv c \pmod{n}$ gives $(c - 1, n)$ a proper factor of n. (Compare 1.4(b) above.)

(c) Check that the following numbers all pass Miller's test to bases 2, 3 and 5:

(i) 14 386 156 093, (ii) 15 579 919 981, (iii) 18 459 366 157,

(iv) 19 887 974 881, (v) 21 276 028 621.

By applying Miller's test to base 7 with n equal to each of (i), (ii) and (iv), find (as in 1.4(b) above), c with $c^2 \equiv 1 \pmod{n}$ and $c \not\equiv \pm 1 \pmod{n}$. [This c is the last residue produced by Miller's test.] By finding $(c - 1, n)$, find a proper factor of n. Does this produce a proper factor for (iii) and (v)?

(d) Here are some more numbers which are strong pseudoprimes to base 2, 3 and 5. Check this and see whether they can be factorized as in (c) above, using base 7 (or possibly 11):

(i) 1 157 839 381, (ii) 3 215 031 751, (iii) 3 697 278 427,

(iv) 5 764 643 587, (v) 6 770 862 367.

Assuming that you *can* factorize (ii), show that it is also a Carmichael number.

(e) Use Miller's test to show that the numbers $k \cdot 10^8 + 1$, $1 \le k \le 12$, are all composite except possibly for $k = 6$ and $k = 7$. (For some k, e.g. $k = 2$, Miller's test is hardly necessary!)

(f) Use Miller's test (or anything else that comes to mind) on the numbers $5\,000\,000$ to $5\,000\,100$ to decide which could possibly be prime. Use Miller's test to find the first number $> 5 \cdot 10^8$ which could possibly be prime.

1.8 Project   Adapt the Miller's test program to measure the number of steps up to the point where the test terminates (by reaching an odd exponent or a residue other than 1 or $n - 1$). Now take the primes one by one from a file (see 1.10 and 1.11 of Chapter 2) and choose a fixed base b, such as 2. Modify the program so that it prints out a prime p for which the number of steps in Miller's test is greater than the number of steps for any previous prime. For example, when $b = 2$, you should find that $p = 2$ gives one step, $p = 3$ gives two steps, $p = 17$ gives three steps (namely, $2^{16} \equiv 1, 2^8 \equiv 1, 2^4 \equiv 16 \equiv -1 \pmod{17}$), $p = 73$ gives four steps, $p = 257$ gives six steps (so the smallest p giving five steps is > 257), $p = 6529$ gives eight steps. As pointed out above, a large number of steps indicates that a much lower power of b than the $(n-1)^{\text{st}}$ is $\equiv 1 \pmod{n}$.

1.9 Project   Find all the strong pseudoprimes to base 2 which are less than $10\,000$. Make sure that there are no primes in your list! (You should find just five strong pseudoprimes.)

1.10 Project   Write a program to show that the smallest number which is both a strong pseudoprime to base 2 and a Carmichael number is $15\,841$. You can assume that all Carmichael numbers are given as in Chapter 4,4.9.

1.11 Project   There is a simple observation which makes the program for Miller's test more elegant. Let the binary decomposition of $n - 1$ be $d_k d_{k-1} \ldots d_1 d_0$, where each d_i is 1 or 0 and $d_k = 1$. Then (i) $n - 1$ is even if and only if $d_0 = 0$, and (ii) when this holds, the binary decomposition of $(n - 1)/2$ is $d_k d_{k-1} \ldots d_1$. Incorporate these into a Miller's test program. (Does the change make the program run any faster? Presumably you need to find the binary decomposition of $n - 1$ *first* now?)

2. Probabilistic primality testing

It was stated in 1.7(b) above that the smallest simultaneous strong pseudoprime to bases 2, 3 and 5 is $25\,326\,001$. Thus (assuming this result) any number less than $25\,326\,001$ can be proved prime or composite by at most three applications of Miller's test. We do not propose to pursue this method of primality proof here, since there are other interesting methods available which do not involve first finding the smallest simultaneous pseudoprime to a number of bases. However, it is worth pointing out that in some sense the more Miller's tests (for different bases) which a number passes, the 'more likely' it is to be prime. In fact the following can be shown (for a proof, see for example [K.M.Rosen (1988), Section 8.4]).

2.1 Theorem *Let n be odd and composite. Then n passes Miller's test for at most $(n-1)/4$ bases b with $1 \leq b \leq n-1$.*

Notice that this says that there is no 'strong' analogue of Carmichael numbers: n can never be a strong pseudoprime to every base. It is not practicable to try as many as $(n-1)/4$ bases when n is large, but we can reason informally as follows. Given an integer n, choose a 'random' base b with $1 \leq b \leq n-1$. Suppose that n passes Miller's test to base b. The probability of having picked one of the bases b for which this happens would be, for a composite n, at most $\frac{1}{4}$, so it is plausible that the probability that n is composite is at most $\frac{1}{4}$. (There are clearly some very bad choices for b, such as 1 or $n-1$. See 1.4(a) above.) Moreover, if we successfully perform Miller's test to k different bases b in the same range and assume that they are 'independent', then the probability that n is composite becomes $1/4^k$. Thus our confidence in asserting that n is prime becomes stronger as k increases. In fact this method, known as *Rabin's probabilistic primality test*, is routinely used to find very large 'probable primes', with maybe n having 50 digits and k being around 100. In Miller's original paper (1976) he showed that, on the assumption that a conjecture called the *generalized Riemann hypothesis* held, a composite number n would fail his test for some base $< 2(\log n)^2$. For example, if n is a composite number around 10^{50} then this implies that around 27 000 Miller's tests are enough to establish compositeness.

2.2 Computing exercises

(a) Adapt the Miller's test program P5_1_5 or P5_1_6 so that it takes say 10 random choices of base b, with $1 \leq b \leq max$ for a number max which can be input with n. Thus you need to add max and a counting variable such as test to the **VAR** statement:

```
max, test:  integer;
```

and the program will read n and max rather than n and b. Insert

```
Randomize;
```

after the `readln` statement and put a `FOR` loop (`test:= 1 to 10`) to repeat the test 10 times. The base is chosen by

$$\texttt{base:= Random(max - 1) + 2;}$$

which is a random integer b satisfying $2 \leq b \leq max$.

Now apply this multiple Miller's test to the examples 1.7(e) and (f) to see whether the 'probable primes' remain so under the more stringent test. You could of course choose $n-1$ (or $n-2$) for max, but in practice it is more sensible to use a number around 100.

(b) Find the 'probable primes' of the form $n = r^4 + 1$, where $1900 \leq r \leq 2000$. (We shall see later, in Chapter 6,3.4, that for numbers n where $n - 1$ can be factorized fairly easily, there is a good method for proving n really *is* prime.)

(c)* It has been conjectured that the number of primes of the form $r^4 + 1$ where $1 \leq r \leq R$, is of the form

$$Q(R) = \lambda \int_2^R \frac{dx}{\log x},$$

for large R, where λ is a constant (and log is to the base e). Assuming that this holds, and using the results of (b), estimate the value of λ.

6

Euler's theorem, orders and primality testing

Euler's theorem is a simple generalization of Fermat's theorem to the case of a nonprime modulus. It was proved by Euler in 1760. For any a with $(a, n) = 1$, it tells us a positive power of a which is congruent to 1 modulo n. Note that the equation $a^k \equiv 1 \pmod{n}$ cannot hold for any $k > 0$ if a is *not* coprime to n, for it implies that any common factor of a and m is also a factor of 1. The question naturally arises, as to the *smallest* positive power of a which is congruent to 1 modulo n: we have already touched on this in Section 1 of Chapter 5. This smallest power is called the *order of p modulo n*, and the concept of order is one of the most powerful and useful we shall meet. It enables us to prove various *primality tests*, that is *sufficient* conditions for a number to be prime. (Passing Miller's test is a *necessary* condition for primality.) We shall use these tests to find primes for which a proof of primality by trial division would be too lengthy. Recall that the trial division program P 2_1_13 uses the file of primes which contains only primes $< 10^5$. Thus we are 'forbidding' primality proofs by trial division for numbers $> 10^{10}$. This gives the flavour of proving really big numbers are prime.

1. Euler's function (the ϕ-function or totient function)

1.1 Definition *Let $n \geq 1$ and let $\phi(n)$ be the number of integers x satisfying $1 \leq x \leq n$ and $(x, n) = 1$. The function ϕ is called* Euler's function, *or the* totient function.

1.2 Examples Clearly $\phi(1) = 1$, $\phi(2) = 1$. For any $n > 2$, the numbers 1 and $n - 1$ will be coprime to n, so $\phi(n) \geq 2$.

If p is prime, then all the numbers $1, 2, \ldots, p - 1$ are coprime to p and so $\phi(p) = p - 1$. Conversely if n is composite and > 1 then it has some divisor d with $1 < d < n$, and such a number d is not coprime to n. Hence $\phi(n) < n - 1$. It follows that $\phi(n) = n - 1$ if and only if n is prime.

For a prime power p^a, the numbers x with $1 \leq x \leq p^a$ and $(x, p^a) = 1$ are precisely the numbers which are not multiples of p. The multiples of p are $p, 2p, 3p, \ldots, p^{a-1}p$, which are p^{a-1} in number. So $\phi(p^a) = p^a - p^{a-1} = p^a(1 - \frac{1}{p})$.

We now seek a general formula for $\phi(n)$. This is provided by the last example of 1.2 and the following result.

1.3 Proposition   *The function ϕ is multiplicative, that is, if $(m, n) = 1$, then $\phi(mn) = \phi(m)\phi(n)$.*

Proof   The proof is a matter of carefully counting the numbers x with $1 \leq x \leq mn$ and $(x, mn) = 1$. Note that $(x, mn) = 1$ if and only if $(x, m) = 1$ and $(x, n) = 1$. Arrange the numbers x in a rectangular array:

1	$m + 1$	$km + 1$	$(n - 1)m + 1$
2	$m + 2$	$km + 2$	$(n - 1)m + 2$
......	....	....	....
r	$m + r$	$km + r$	$(n - 1)m + r$
......	....	....	....
m	$2m$	$(k + 1)m$	nm

Clearly, if $(r, m) = d > 1$ then no number in the rth row is coprime to mn, since $d \mid (km + r)$ and $d \mid mn$. So consider the rows with $(r, m) = 1$; note that there are $\phi(m)$ of them. We claim that, in such a row, all entries are coprime to m and exactly $\phi(n)$ entries are coprime to n. The first statement follows because $(km + r, m) = (r, m) = 1$. For the second, note that there are n entries in the row and no two are congruent modulo n. [For $km + r \equiv k'm + r \pmod{n}$ implies $km \equiv k'm \pmod{n}$ and, *since m and n are coprime*, this implies $k \equiv k' \pmod{n}$ and hence $k = k'$.] Hence, mod n, the rth row consists of $0, 1, 2, \ldots, n - 1$ in some order, and exactly $\phi(n)$ of these are coprime to n.

The theorem now follows, since the $\phi(n)$ entries in row r, being coprime to both m and n, will be precisely the entries in that row coprime to mn. □

It is now an easy matter to write down a general formula for $\phi(n)$, using the last example of 1.2 and 1.3.

1.4 Corollary   *If $n = p_1^{n_1} p_2^{n_2} \ldots p_k^{n_k}$ is the prime-power decomposition of n (so that the p_i are distinct primes and each n_i is ≥ 1), then*

$$\phi(n) = p_1^{n_1} \left(1 - \tfrac{1}{p_1}\right) p_2^{n_2} \left(1 - \tfrac{1}{p_2}\right) \ldots p_k^{n_k} \left(1 - \tfrac{1}{p_k}\right)$$

$$= n \left(1 - \tfrac{1}{p_1}\right) \left(1 - \tfrac{1}{p_2}\right) \ldots \left(1 - \tfrac{1}{p_k}\right).$$

Proof   The various prime powers are coprime so we apply 1.3.

1.5 Exercises

(a) Make a table of values of $\phi(p^a)$ for small primes p and integers $a \geq 1$, and, using the multiplicative property 1.3, find all values of n for which $\phi(n) = 6$. (A more ambitious exercise is to find all n with $\phi(n) = 48$: this has no fewer than 11 solutions.) Show also that there is no n for which $\phi(n) = 14$. A useful observation here is that, if p is prime and $p \mid n$, then $(p-1) \mid \phi(n)$. This rules out quite a lot of prime divisors of n. (Because 14 is not a value of $\phi(n)$, it is sometimes called a *non-totient*. In fact (a harder exercise) ϕ never takes any value of the form $2 \cdot 7^k$, $k \geq 1$. See also 1.15 below.)

(b) Find the prime-power decomposition of 10! and hence find $\phi(10!)$.

(c) Show that, for any $n > 2, \phi(n)$ is even. [Hint. Either $n = 2^k$ for some $k > 1$ or $p \mid n$ for some odd prime p. This can also be seen by noting that for $n > 2$ the numbers x coprime to n and satisfying $1 \leq x \leq n$ can be put into pairs $\{x, n - x\}$ of *distinct* numbers.]

(d) By writing $n = 2^k r$, where r is odd, show that, if $\phi(n) = n/2$, then n is a power of 2.

(e) Use the formula for $\phi(n)$ to describe all n for which $\phi(n)$ is a multiple of 4. (For example, if $8 \mid n$, then $4 \mid \phi(n)$.)

(f) Suppose that n is composite. Show that $\phi(n) \leq n - \sqrt{n}$. [Hint. If n is composite then it has a prime factor $p \leq \sqrt{n}$. Which multiples of p will be $\leq n$?]

(g)* Show that, if $\phi(n) = 2 \cdot 3^{6k+1}$, where $k \geq 2$, then $n = 3^{6k+2}$ or $2 \cdot 3^{6k+2}$. [It is not hard to check that, for these values of $n, \phi(n)$ has the stated value. The hard thing is to show the converse.] It follows from this that there are infinitely many values of y for which the equation $\phi(x) = y$ has *exactly two* solutions for x. Amazingly, it is not known whether there are *any* values of y for which the equation $\phi(x) = y$ has *exactly one* solution for x. (Compare 1.10 below.)

(h) Verify that, if $n = 1$ or $n = 2^i$ or $n = 2^i 3^j$ for integers $i \geq 1$, $j \geq 1$, then $\phi(n)$ divides n.

(i) This is the converse of (h), and is slightly harder. In fact, assume $\phi(n) \mid n$ and $n > 1$. Let P be the product of all distinct primes p dividing n and Q be the product of factors $p - 1$ for all such p. Show that $Q \mid P$ and deduce that Q is not divisible by any square > 1, and that P can only be the product of one or of two primes. Deduce from this that no prime $p > 3$ can be a factor of P and hence that in fact n has one of the values given in (h) above.

(j) As a contrast to 1.3, show that, if $m \mid n$, then $\phi(mn) = m\phi(n)$. (Use 1.4.)

(k) Show that if n is *odd* and $\phi(n) = 2^r$ for some $r \geq 1$, then n must be a product of *distinct* primes (or indeed prime itself). Using the fact that $2^{32} + 1$ is *not* prime (compare Chapter 2, 1.16(c)), show that there is no odd number n with $\phi(n) = 2^{32}$. On the other hand, for $1 \leq r \leq 31$, there do exist odd numbers n with $\phi(n) = 2^r$. What are the *even* numbers n with $\phi(n) = 2^{32}$? [It appears that the smallest number y with the property that $\phi(n) = y$ has an even solution for n but no odd solution is $y = 2^9 \cdot 257^2 = 33\,817\,088$. See *Amer. Math. Monthly* **98** (1991), p.443.]

1.6 Remark: ϕ and combinatorics   There is a very curious connexion between Euler's function and combinatorics. If we consider circular necklaces made by stringing coloured beads on a circular wire, then two such necklaces can be considered *equivalent* if the beads on one can be rotated round the wire so that their colours exactly match the beads of the other. (We do not allow turning the wire over.) If there are k colours available then it can be shown that the number of inequivalent necklaces is

$$\frac{1}{n} \sum_{d|n} \phi\left(\tfrac{n}{d}\right) k^d,$$

the sum being over all divisors d of n.(This is proved in books on discrete mathematics, for example [Graham *et al.* (1989)]. It is also very entertainingly covered in [Knuth *et al.* (1989), pp.62–65].) As a special case, you can verify that when n is a prime p, the formula reduces to $\frac{1}{p}(k^p - k + pk)$. The fact that this must of course be an *integer* implies that $p \mid (k^p - k)$, which is Fermat's theorem. [Some observations such as these appear to have been made as long ago as 1910 by A. Thue who deduced both Fermat's and Euler's theorems by combinatorial arguments.]

The case $k = 1$ of the above remark is worth noting separately, and since we won't prove the general result we shall prove this special case.

1.7 Proposition   *For any $n \geq 1$, $n = \sum_{d|n} \phi(d)$, the sum being over all (positive) divisors d of n.*

Proof   Let $C_d = \{x : d \leq x \leq n$ and $(x, n) = d\}$. Clearly no two sets C_d and $C_{d'}$, for distinct divisors d and d', can overlap, and every x with

$1 \le x \le n$ lies in some C_d. Thus $n = \sum \#(C_d)$, summed over all $d \mid n$. (Here, $\#$ just means the number of elements in a set.) Now

$$x \in C_d \Leftrightarrow d \mid x, \ 1 \le \frac{x}{d} \le \frac{n}{d} \text{ and } (\tfrac{x}{d}, \tfrac{n}{d}) = 1.$$

Consequently, C_d is in one-to-one correspondence with the set of integers y satisfying $1 \le y \le n/d$ and $(y, n/d) = 1$, so that $\#(C_d) = \phi(n/d)$. But as d runs through the divisors of n so does n/d, so that

$$n = \sum_{d \mid n} \phi(n/d) = \sum_{d \mid n} \phi(d).$$

$\square$

1.8 Computing exercise   Write a program, based perhaps on the trial division program P2_1_13, to calculate $\phi(n)$ for a given n. There are several ways to do this, one of which is based on the formula of 1.4. That is, ϕ is initialized to 1, and the trial division proceeds as before. For each prime p dividing n, when the highest power $x = p^a$ dividing n has been found, the value of ϕ is multiplied by $x(p-1)/p$. Remember that you only want to do this last step when in fact p *is* one of the primes which has been found to divide n. (Alternatively initialise ϕ to n and multiply by $(p-1)/p$ *once* for every prime p dividing n.) Verify that your program works by checking small values of n; also check that $\phi(5186) = \phi(5187) = \phi(5188) = 2592$. (A rather rare example!) Another rarity is $\phi(25\,930 + 5k) = 2^7 \cdot 3^4$ for $k = 0, 1, 2$. In fact there is another number not much bigger than $25\,940$ which also has the same value for ϕ; can you find it?

1.9 Project   Use a program for calculating ϕ (as in 1.8) to calculate, for various values of n, the *average value* of $\phi(k)/k$ for $k = 1, 2, \dots, n$. (This is the sum of the values, divided by n.) There is a theoretical estimate for this average, of the form $6/\pi^2 + M(\log n/n)$, where M is constant (and log is to the base e). Estimate the value of M. [See Schroeder (1986), p. 133.]

1.10 Exercise: $n\phi(n)$ determines n   From 1.5(a) there are often many n for which $\phi(n)$ has a given value. It is curious that, however, if $n\phi(n) = m\phi(m)$, then necessarily $n = m$. Here is a suggestion for proving this. First prove from the formula for $\phi(n)$ that, if p is the *largest* prime dividing $n\phi(n)$, then p is also the largest prime dividing n. Supposing that $n\phi(n) = m\phi(m)$, deduce that p is also the largest prime dividing m, and further that the power of p which divides each side of the equation is *odd*, say p^{2r-1}, where p^r is the power of p which divides both m and n. Thus $n = p^r N$ and $m = p^r M$, say, where $p \nmid N$ and $p \nmid M$. Now use the given equation to deduce that $N\phi(N) = M\phi(M)$.

This reduces the number of distinct prime factors by one and eventually reduces the problem to proving that if $n = p^a$ and $m = q^b$ for primes p and q, then $n\phi(n) = m\phi(m)$ implies $p = q$ and $a = b$, that is $n = m$. (Why is it not possible to run out of prime factors on one side of the equation before you run out of them on the other side?)

Determine the unique values of n for which $n\phi(n)$ has the following values: 20, 12, 42, 40, 500, 1000, 10 100, 100 000, 9 003 000.

1.11 Project   Write a program which, given an integer $k > 0$, either shows that there is no n with $n\phi(n) = k$, or else finds the unique n with this property, using 1.10. Thus you will start by finding the largest prime factor p of k, and determining the highest power of p dividing k. If this power is even then you can deduce no such n exists. If the power is p^{2r-1}, then you note a factor p^r in n and replace k by $k/p^r\phi(p^r)$. Of course, if the latter is not an integer, then no such n exists.

1.12 Project   It is not known whether there are infinitely many n for which $\phi(n) = \phi(n+1)$. Find all the values of $n < 10\,000$ for which this holds. Find also the values of $n < 10\,000$ for which $\phi(n) = \phi(n+3)$. (There are rather few of these!)

1.13 Project: iterating the Euler function   Show that forming a sequence $n, \phi(n), \phi^2(n) = \phi(\phi(n)), \phi^3(n) = \phi(\phi(\phi(n))), \ldots$, the value always eventually becomes 2. For $n > 2$, let the *class* $C(n)$ of n be the value of k for which $\phi^k(n) = 2$. Calculate the exact values of $C(2^r)$ and $C(2 \cdot 3^r)$. Now, for a given value of k, we can ask for the smallest value of n which has class k. Write a program to show that, for $k = 1, 2, 3, 4, 5, 6$ this smallest n is (respectively) $n = 3, 5, 11, 17, 41, 83$. How does the sequence of smallest n continue? Does it contain only primes? Some more properties of the class are given in the next exercises.

1.14 Exercises: some properties of the class as in 1.13
These come from [Shapiro (1943)]. Some of them depend on the observation that, while $\phi(mn) = \phi(m)\phi(n)$ when $(m, n) = 1$, we have $\phi(mn) = m\phi(n)$ when $m \mid n$. This follows from the formula 1.4 above and has already been noted in 1.5(j). Remember also that $\phi(n)$ is even when $n > 2$ (see 1.5(c) above). We assume $n > 2$.

(a) When n is odd, so $(2, n) = 1$, we have $\phi(2n) = \phi(n)$. Deduce that $\phi^k(2n) = \phi^k(n)$ for all k and hence that $C(2n) = C(n)$.

(b) When n is even, so $2 \mid n$, we have $\phi(2n) = 2\phi(n)$ and, since $\phi(n)$ is even, we have $\phi^2(2n) = 2\phi^2(n)$. Deduce that $\phi^k(2n) = 2\phi^k(n)$ so long as $\phi^{k-1}(n)$ is even, and hence for $k \le C(n)$. Deduce that $\phi^{C(n)}(2n) = 4$ and $C(2n) = C(n) + 1$.

(c) Use (a) and (b) to show that, if n is odd, then $C(2^r n) = C(2^{r-1}n)+1 = \ldots = C(2n) + r - 1 = C(n) + r - 1$. Note that this implies $C(2^r n) = C(n) + C(2^r)$.

(d) Use (a) and (b) to show that, if n is even, then $C(2^r n) = C(2^{r-1}n) + 1 = \ldots = C(n) + r$. Note that here $C(2^r n) = C(n) + C(2^r) + 1$.

(e) We have $\phi(3n) = 3\phi(n)$ or $2\phi(n)$ according as $3 \nmid n$ or $3 \mid n$. Use this to show that, for $k \leq C(n)$, $\phi^k(3n) = 2\phi^k(n)$ or $3\phi^k(n)$. Deduce that $\phi^{C(n)}(3n) = 4$ or 6 and that $C(3n) = C(n) + 1 (= C(n) + C(3))$.

Shapiro goes on to deduce that $C(pn) = C(n) + C(p)$ for any odd prime p. Perhaps you would like to try this for $p = 5$. From the above results it now follows that $C(mn) = C(m) + C(n)$ unless both m and n are even, in which case we add 1 to the right-hand side. (It is convenient to define $C(2) = 0$ to cover all cases.) If you have experimental data from 1.13 above then you can check this against your results.

1.15 (Computing?) exercise Determine all the numbers $n \leq 100$ (or ≤ 1000 if you can) which are *nontotients*, that is, which are not equal to $\phi(n)$ for any n. Is there a better method than making a table of values of $\phi(p^a)$ for $p^a \leq 101$ (or ≤ 1001) and working out all possible products of such values, for distinct primes, which are ≤ 100 (or 1000)? Compare 1.5(a). Of course all odd numbers > 1 are non totients (compare 1.5(c)). The even non totients ≤ 200 are 14, 26, 34, 38, 50, 62, 68, 74, 76, 86, 90, 94, 98, 114, 118, 122, 124, 134, 142, 146, 152, 154, 158, 170, 174, 182, 186, 188, 194.

1.16 Project If we change the minus signs to $+$ in the formula for ϕ (1.4) we obtain a new function:

$$\overline{\phi}(n) = n\left(1 + \tfrac{1}{p_1}\right)\left(1 + \tfrac{1}{p_2}\right)\ldots\left(1 + \tfrac{1}{p_k}\right) = \prod p_i^{a_i - 1}(p_i + 1).$$

where as before the p_i are the distinct primes dividing n. Use the fact that, for $p_i > 2$, the prime factors of $p_i + 1$ are 2 and primes $< p_i$, to show that iteration of $\overline{\phi}$ (i.e. $n, \overline{\phi}(n), \overline{\phi}(\overline{\phi}(n)), \ldots$) gives eventually a number of the form $2^a 3^b$, which then becomes $2^{a+1}3^b, 2^{a+2}3^b$, etc. Write a program which calculates $\overline{\phi}$ and the number of steps taken to reach the form $2^a 3^b$. Why is the number of steps the same for $n, 2n$ and $3n$ $(n \geq 1)$?

Test the following conjecture: if n requires more steps to reach the form $2^a 3^b$ than is required by all numbers $< n$, then n is prime. (The first n to require 1,2,3,4,5,6,7 steps are 5, 13, 37, 73, 673, 1993, 15013 respectively.)

1.17 Project It was proved by R.E. Dressler in 1970 that if

$$N(x) = \#\{n > 0: \phi(n) \leq x\}$$

then $N(x)/x \to A$ as $x \to \infty$, where A is a certain constant. Estimate the value of A. [The theoretical value of A is approximately 1.94.]

2. Euler's theorem and the concept of order

2.1 Euler's theorem (1760)   *Let $n > 0$ and suppose $(a, n) = 1$. Then $a^{\phi(n)} \equiv 1 \pmod{n}$.*

Proof   Let $r_1, r_2, \ldots, r_k$ $(k = \phi(n))$ be the integers ≥ 1 and $\leq n$ which are coprime to n. Consider $ar_1, ar_2, \ldots, ar_k$. No two of these are congruent mod n $(ar_i \equiv ar_j$ implies $r_i \equiv r_j \pmod{n}$ since $(a, n) = 1)$ and they are all coprime to n $((a, n) = (r_i, n) = 1$ implies $(ar_i, n) = 1)$. Since there are $\phi(n)$ of them their least positive residues mod n must be precisely the numbers $r_1, r_2, \ldots, r_k$ again, in some order. Hence

$$r_1 r_2 \ldots r_k \equiv ar_1 ar_2 \ldots ar_k \pmod{n},$$

so that, cancelling the r_i because they are coprime to n, we have the required congruence $1 \equiv a^k \pmod{n}$.   □

2.2 Computing exercise   Since $a \cdot a^{\phi(n)-1} \equiv 1 \pmod{n}$, it follows that $a^{\phi(n)-1}$ is the inverse of a, mod n. Use this and a program for ϕ as in 1.8 above to write a program for solving linear congruences $ax \equiv b \pmod{n}$, where $(a, n) = 1$. How would you change this to solve such congruences without the assumption $(a, n) = 1$?

2.3 Exercises
(a) Use Euler's theorem to show that $a^{40} \equiv 1 \pmod{100}$ provided $(a, 100) = 1$. Deduce that $7^{400} - 3^{400}$ is divisible by 100. Is it divisible by 1000? Note that using $100 = 2^2 5^2$ and Euler's theorem separately on 4 and 25 you can deduce that $a^{20} \equiv 1 \pmod{100}$ for $(a, 100) = 1$, since the congruence holds mod 4 and mod 25.

(b) Show that if $(2, a) = (2, b) = 1$ and $(5, a) = (5, b) = 1$ then $a^{1000} - b^{1000}$ is divisible by 10 000.

(c) Show that $10^{6(p-1)} \equiv 1 \pmod{9p}$ for p a prime > 5. Deduce that there are infinitely many integers n for which p divides the 'repunit' $11 \ldots 1$ consisting of n ones in decimal notation. Is the same true for $p = 3$?

(d) Let $n = rs$ where $r > 2$, $s > 2$ and $(r, s) = 1$, and let $(a, n) = 1$. Recall from 1.5(c) above that $\phi(n)$ is even since $n > 2$. Use Euler's theorem to show that

$$a^{\phi(r)\phi(s)/2} \equiv 1 \pmod{n}.$$

[Hint. Show this mod r and mod s.] In view of the multiplicativity of ϕ (1.3) this shows that $a^{\phi(n)/2} \equiv 1 \pmod{n}$.

(e) Let $\alpha = r\pi/s$ where $(r, s) = 1$. Write $s = 2^a b$, where $a \geq 0$ and b is odd, and write $\beta = 2^a \alpha = r\pi/b$. Use Euler's theorem, $2^k \equiv 1 \pmod{b}$ for $k = \phi(b)$, to show that $\cos(2^{k+1}\beta) = \cos(2\beta)$. Deduce that the sequence

$$\cos(\alpha), \cos(2\alpha), \cos(2^2\alpha), \cos(2^3\alpha), \ldots$$

has only a finite number of different terms.

The result of 2.3(a) above shows that we shall often have a *smaller* power of a than the power $\phi(n)$ which is congruent to 1 mod n. In fact it shows that if n is divisible by two distinct odd primes then this will *always* happen, for *any* a coprime to n. We shall have a good deal of use to make of the following concept.

2.4 Definition   *Let $(a, n) = 1$. The smallest number $k > 0$ such that $a^k \equiv 1 \pmod{n}$ is called the* order *of a mod n. It is written $k = \mathrm{ord}_n\, a$, and we also say that a has order k mod n. Of course $\mathrm{ord}_n a \leq \phi(n)$, by 2.1. Note that if $(a, n) > 1$ and $k > 0$ then it is impossible for a^k to be congruent to 1 mod n. If a has order $k \pmod{n}$ then some books use the expression 'a belongs to the exponent $k\pmod{n}$'.*

2.5 Proposition   *Suppose $(a, n) = 1$. Then $a^k \equiv 1 \pmod{n}$ if and only if $\mathrm{ord}_n a \mid k$. Thus the powers of a which are congruent to 1 are not merely $\geq \mathrm{ord}_n a$ but precisely the multiples of this order. In particular, $\mathrm{ord}_n a \mid \phi(n)$ and, if p is prime, then $\mathrm{ord}_p a \mid (p-1)$.*

Proof   Suppose $a^k \equiv 1$ where $k > 0$ and let $k = q\,\mathrm{ord}_n a + r$, where $0 \leq r < \mathrm{ord}_n a$. Then, mod n, $1 \equiv a^k = \left(a^{\mathrm{ord}_n a}\right)^q a^r \equiv a^r$. Hence $r = 0$ by definition of the order as the smallest positive power of a which is congruent to 1, and this proves the 'only if' part of the proposition. The 'if' part is clear from the definition of order.   □

2.6 Exercises

(a) Let $n = 12$, so that $\phi(n) = 4$. According to 2.5, the order mod 12 of any number a with $(a, 12) = 1$ must be 1, 2 or 4. Find all the possible orders. Does any equal 4? (Compare 2.3(d).)

(b) Show that the order of 2 mod 11 is $\phi(11) = 10$. To do this, it is enough to show that $2^k \not\equiv 1 \pmod{11}$ for $k = 2$ and 5, these being the proper factors of 10. What is the order of 2 mod 22? Show that 7 and 13 both have order $\phi(22) = 10$, mod 22.

(c) Show that there is no number a with $\mathrm{ord}_8 a = \phi(8) = 4$. Is this predicted by 2.3(d)? Does there exist a with $\mathrm{ord}_4 a = \phi(4) = 2$? Find all the numbers a with $\mathrm{ord}_7 a = \phi(7) = 6$, and all the numbers b with $\mathrm{ord}_{14} b = \phi(14) = 6$.

(d) Let p be prime and suppose that a has order 3 mod p. (Note that this implies that $3 \mid (p - 1)$ and since p has to be odd it follows that $6 \mid (p - 1)$.) Show that $a^2 + a + 1 \equiv 0 \pmod{p}$ and deduce that $(a+1)^6 \equiv 1 \pmod{p}$. Show that in fact $a+1$ has order exactly 6 mod p.

(e) Let p be prime and suppose that a has order n mod p^k and order m mod p^{k-1}, where $k \geq 2$. Show that $n = m$ or $n = mp$. [First show that $a^n \equiv 1 \pmod{p^{k-1}}$ and deduce that $m \mid n$. Then use $a^m = 1 + \lambda p^{k-1}$ for some integer λ, and the binomial theorem, to show that $a^{mp} \equiv 1 \pmod{p^k}$. Deduce from this that $n \mid mp$. The result follows from $m \mid n$ and $n \mid mp$.] This result is used in Chapter 8,2.6 to prove results about the existence of 'primitive roots'.

(f) Let p be prime and suppose that a has order n mod p^2 and m mod p. Show that $m = n$ if and only if $a^{p-1} \equiv 1 \pmod{p^2}$. [The point here is that, if $a^{p-1} \equiv 1 \pmod{p^2}$, then $n \mid (p - 1)$; now use the result of (e), with $k = 2$.] Later (1.3(a) in Chapter 8) we shall see how, in principle at least, we can find the values of a which satisfy this congruence, for a given value of p. Fixing a, e.g. $a = 2$, it is rather rare for a prime p to satisfy the congruence (see 2.8 below).

(g) Suppose that p is prime and $a^{p-1} \equiv 1 \pmod{p^2}$ and $a^m \equiv 1 \pmod{p}$. Show that $a^m \equiv 1 \pmod{p^2}$. [Let $r = \text{ord}_p a$; write $p - 1 = rk$ and $m = rh$, $a^r = 1 + cp$. Use the first hypothesis to show $p \mid c$ and then write $a^m = (1 + cp)^h$ and use the binomial theorem.]

(h) Suppose that p is a prime and $p \mid F_k$ where $F_k = 2^{2^k} + 1$ is the kth Fermat number. Show that $\text{ord}_p 2 = 2^{k+1}$ and deduce that $2^{k+1} \mid (p-1)$. For an improvement, see Chapter 11,2.3(d).

(i) Let $(m, n) = (x, m) = (x, n) = 1$ and let $a = \text{ord}_m x$, $b = \text{ord}_n x$. Show that $x^k \equiv 1 \pmod{mn}$ if and only if $a \mid k$ and $b \mid k$. Deduce that $\text{ord}_{mn} x = \text{lcm}(a, b)$.

(j) Let $a = \text{ord}_m x$, and let $s \geq 1$. Show that $\text{ord}_m(x^s) = \frac{a}{(a,s)}$. [This amounts to showing that $\frac{a}{(a,s)} \mid t \Leftrightarrow (x^s)^t = 1$.]

2.7 Definition *A number a with $(a, n) = 1$ and $\text{ord}_n a = \phi(n)$ is called a primitive root mod n.* Thus according to 2.3(d) if n is divisible by two distinct odd primes it has no primitive root. Likewise from 2.6(c) 8 has no primitive root. For p prime, the condition for a to be a primitive root mod p is that $\text{ord}_p a = p - 1$. We shall have more to say about primitive roots in Chapter 8.

2.8 Project Verify that $a^{p-1} \equiv 1 \pmod{p^2}$ when $a = 5$ and p is any of the primes $20\,771, 40\,487, 53\,471\,161$. Find the smallest prime p for which this happens (a) when $a = 3$, (b) when $a = 2$ and (c) when $a = 11$. Find other solutions for $a < 50$. [There is an extraordinary connexion with 'Fermat's last theorem', due to A. Wieferich in 1909. If

$x^p + y^p = z^p$ where p is an odd prime which does not divide any of the positive integers x, y, z, then $2^{p-1} \equiv 1 \pmod{p^2}$. Of course the 'last theorem' states that there are no solutions for any $p > 2$.]

2.9 Exercise Let p and q be distinct primes and suppose that r is a prime factor of both

$a = (p^q - 1)/(p - 1) = p^{q-1} + p^{q-2} + \ldots + 1$ and
$b = (q^p - 1)/(q - 1) = q^{p-1} + q^{p-2} + \ldots + 1$.

(a) Show that if $r \mid (p - 1)$ then $a \equiv q \pmod{r}$ and deduce $r = q$. Since $q \nmid b$ this shows that $r \nmid (p - 1)$ and similarly $r \nmid (q - 1)$.

(b) Show that if $p = 2$ then $2^q - 1 \equiv q + 1 \equiv 0 \pmod{r}$. Deduce that $\mathrm{ord}_r 2 = q$ and hence $q \mid (r - 1)$ and $r \mid (q + 1)$. Deduce that $p > 2$ and similarly $q > 2$.

(c) Show that a and b are odd and deduce that r is odd.

(d) Use $p^q \equiv 1 \pmod{r}$ and (a) to show that $\mathrm{ord}_r p = q$ and deduce $q \mid (r - 1)$. Similarly $p \mid (r - 1)$. Deduce that r has the form $2kpq + 1$.

(e) Take $p = 17$ and write a program to find the smallest prime q for which $r = 2pq + 1$ is a common factor of a and b as above. [This example has occurred before, in Chapter 4 !]

2.10 Exercise: least common multiple of orders   Given two elements a and b of orders respectively r and s $\pmod{n}$, there is a simple construction for finding another element c of order $\mathrm{lcm}(r, s)$, as follows. Using the result of Chapter 1, 2.6(b), there exist numbers u, v, x, y such that $r = ux$, $s = vy$ and (among other things) $(u, v) = 1$ and $uv = \mathrm{lcm}(r, s)$. Define $c \equiv a^x b^y \pmod{n}$. The claim is that this has order uv. First show that $c^{uv} \equiv 1 \pmod{n}$ so that the order is a factor of uv. Then assume $c^k \equiv a^{kx} b^{ky} \equiv 1 \pmod{n}$. Raise each side to the power u and deduce that $s \mid kuy$ and hence $v \mid k$; similarly show $u \mid k$. Deduce that uv is a factor of the order.

2.11 Project: implementing the previous exercise   Write a program to calculate c from a, b and n. (Compare Chapter 2,1.17.) Remember that a and b must be coprime to n. This program will come in useful later in Chapter 8, 2.11, so it is worth incorporating program P2_1_17 as a procedure (accepting inputs r and s and delivering outputs x and y), and also perhaps using the PowerRule procedure as in P4_2_3, in order to calculate a^x and b^y. For some prime values $n = p$, see if you can, by taking say $a = 2$ and making a cunning choice of b, find an element of the largest possible order $p - 1$. For $n = 100$, what is the largest order you can achieve?

2.12 Exercise: Sierpinski–Selfridge formula giving only composites   In 1960, Sierpinski showed that there are infinitely many k for which $N = k \cdot 2^n + 1$ is composite for *all* $n \geq 1$. In 1963 Selfridge showed that $k = 78\,557$ has this property, and here is your chance to show the same thing. In fact, you can show that N is always divisible by 3, 5, 7, 13, 19, 37 or 73. For example, consider 7: when is $N \equiv 0 \pmod 7$? Since $78\,557 \equiv 3 \pmod 7$ we are solving $2^{n-1} \equiv 1 \pmod 7$. A short calculation shows that $\mathrm{ord}_7 2 = 3$, so by 2.5 this is equivalent to $3 \mid (n-1)$, i.e. $n \equiv 1 \pmod 3$. Similarly consider 13: when is $N \equiv 0 \pmod{13}$? Since $78\,557 \equiv -2 \pmod{13}$ we are solving $2^{n+1} \equiv 1 \pmod{13}$. We find that $\mathrm{ord}_{13} 2 = 12$, so using 2.5 we get $12 \mid (n+1)$, i.e. $n \equiv -1 \pmod{12}$. Show in a similar way that $3 \mid N \iff n \equiv 0 \pmod 2$; $5 \mid N \iff n \equiv 1 \pmod 4$; $19 \mid N \iff n \equiv -3 \pmod{18}$; $37 \mid N \iff n \equiv -9 \pmod{36}$ and $73 \mid N \iff n \equiv 3 \pmod 9$. Now check that every n is covered by at least one of the seven congruences (work mod 36 for this). (Compare Chapter 3,1.6.)

2.13 Project: closed polygons   The idea for this project comes from [Abelson and diSessa (1980), pp.123,128.] We begin with a theoretical calculation. Let p be an odd prime, $k \geq 2$ and let n, m be the orders of 2 mod p^k and mod p^{k-1} respectively (compare 2.6(e)). *We suppose* $n \neq m$.

(a) We claim that, for any $t \geq 0$, $2^t + p^{k-1}$ is, mod p^k, a power of 2. This shows in fact that addition of p^{k-1} *permutes* the powers of 2, mod p^k [why?]. To prove the claim, use the following hints. Write $2^m = 1 + hp^{k-1}$ and raise each side to the power s, say, by the binomial theorem. Use $n \neq m$ to show that $p \nmid h$ and deduce that, for a suitable s, $2^{ms} \equiv 1 + p^{k-1} \pmod{p^k}$. This proves the claim for $t = 0$. Now raise each side of the last congruence to the power r. Show that there exists r such that $2^{msr+t} \equiv 2^t + p^{k-1} \pmod{p^k}$. This proves the general claim.

(b) Let α be the angle $2\pi r/p^k$ (same p and k as before but a new r) where $p \nmid r$. Let $\mathbf{v}_0$ be any nonzero vector and let $\mathbf{v}_s$ be $\mathbf{v}_0$ rotated anticlockwise through an angle $s\alpha$ for any integer s. We consider vectors $\mathbf{w}_0, \mathbf{w}_1, \mathbf{w}_2, \ldots$ where $\mathbf{w}_0 = \mathbf{v}_0$, $\mathbf{w}_1 = (\mathbf{w}_0$ turned through $\alpha) = \mathbf{v}_1$, $\mathbf{w}_2 = (\mathbf{w}_1$ turned through $2\alpha) = \mathbf{v}_3$, $\mathbf{w}_3 = (\mathbf{w}_2$ turned through $4\alpha) = \mathbf{v}_7$, etc. Generally, $\mathbf{w}_i = \mathbf{v}_j$ where $j = 2^i - 1$. The vectors $\mathbf{w}_0, \mathbf{w}_1, \mathbf{w}_2, \ldots$ can be placed end to end to form a polygonal line (see Fig. 6.1): the (external) angle between two edges is double the previous angle between two edges.

Show that, if n is the order of 2 mod p^k, then $\mathbf{w}_n = \mathbf{w}_0$ (as vectors, i.e. they are parallel). The object is to prove that, in the notation of (a), with $m \neq n$, the polygon *closes up* after n sides have been drawn, i.e.

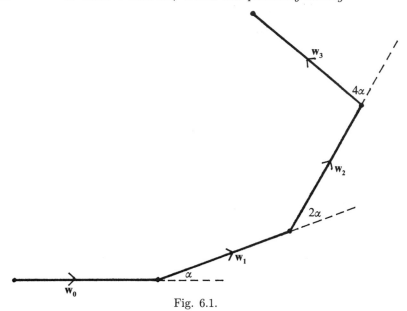

Fig. 6.1.

that $\mathbf{W} = \mathbf{w}_0 + \mathbf{w}_1 + \ldots + \mathbf{w}_{n-1} = \mathbf{0}$. Write ρ_k for rotation of vectors through $k\alpha$ and show, using (a), that $\rho_{c+1}\mathbf{W} = \rho\mathbf{W}$, where $c = p^{k-1}$. Deduce that $\mathbf{W} = 0$, as claimed.

(c) If you know how to write graphics programs, try drawing the polygon for various values of p and k, to test the result of (b) and also to try to judge whether, in the rare cases when $m = n$, the polygon fails to be closed.

3. Primality tests

The main application we shall make here of the idea of order is to testing certain special numbers for primality. Generally speaking, simple tests for the primality of a number n depend on our ability to factorize some number closely associated with n, such as $n \pm 1$. Here is a sample, which gives a converse of Fermat's theorem in a very special case.

3.1 Theorem (primality of $2p + 1$) *Suppose p is prime and let $n = 2p + 1$. If $2^{n-1} \equiv 1 \pmod{n}$ and $3 \nmid n$, then n is prime.*

Proof Let q be prime, $q \mid n$. Since n is odd, we have q odd. The first claim is that $p \mid (q - 1)$. To prove this, note that $4^p = 2^{n-1} \equiv 1 \pmod{n}$ and hence mod q). Hence $\operatorname{ord}_q 2$ is a factor of p, and hence is 1 or p. But if the order is 1 then $4 \equiv 1 \pmod{q}$, which implies $q = 3$ and hence $3 \mid n$, contrary to hypothesis. Hence the order is p, and this shows $p \mid (q-1)$ (compare 2.5 above).

We now show that in fact $n = q$. For $q-1 > p$, so $q \geq p+1 > n/2 \geq \sqrt{n}$ since $n \geq 4$. Thus *every* prime factor q of n is $> \sqrt{n}$, and this is impossible unless n is itself prime. □

3.2 Computing exercise: Cunningham chains of primes Check that $p_1 = 1\,122\,659$ is prime by trial division. How many terms in the sequence

$$p_1, \ p_2 = 2p_1 + 1, \ p_3 = 2p_2 + 1, \ldots$$

are prime? You could of course check primality of *these* by trial division, but try using 3.1 instead. Such chains of primes are called *Cunningham chains* in [Guy (1981), §A7]. There are two other long chains beginning $2\,164\,229$ and $2\,329\,469$. See 3.3(b) below for a small generalization of these chains.

3.3 Exercises

(a) Use a similar method to that of 3.1 to show that, if p is prime and $2^p \equiv 1 \pmod{2p+1}$ then $2p+1$ is prime. How is this result related to 3.1? Is there a similar result with $a^{p-1} \equiv 1 \pmod{2p+1}$ for some other 'base' a?

(b) Let p be an odd prime and $n = ap+1$ where a is even, $2 \leq a < \sqrt{p}$. Suppose that $2^{n-1} \equiv 1 \pmod{n}$ and that $(2^a - 1, n) = 1$. Show that, for any prime q dividing n, the order of $2^a \bmod q$ is p, and deduce $p \mid (q-1)$. Now show that $q > \sqrt{n}$ and deduce that n is prime. This result can be used to look for chains of primes of the form $p_1, \ p_2 = ap_1 + 1$, $p_3 = ap_2 + 1, \ldots$.

The theorem 3.1 is of the following kind: Suppose n satisfies the conclusion of Fermat's theorem, that is, n is either a prime or a pseudoprime to some base a. Suppose also that $a^{(n-1)/q} \not\equiv 1$ for a divisor q of $n-1$ (in the case of 3.1, $q = p$ so $(n-1)/q = 2$). Then n is prime. The most useful general result in this direction is the following, which was given by E. Lucas in a slightly weaker form in 1891 but proved by D.H. Lehmer in 1927 (this paper [Lehmer (1927)] makes pleasant reading).

3.4 Theorem (the $n-1$ over q primality test or Lucas's test) *Suppose that $a^{n-1} \equiv 1 \pmod{n}$ but, for each prime q dividing $n-1$, we have $a^{(n-1)/q} \not\equiv 1 \pmod{n}$. Then n is prime.*

Proof The first hypothesis implies that $r = \mathrm{ord}_n a$ satisfies $r \mid (n-1)$. Write $n-1 = kr$, where we suppose for a contradiction that $k > 1$. Let q be a prime dividing k. Then $q \mid (n-1)$ and

$$a^{(n-1)/q} = (a^r)^{k/q} \equiv 1^{k/q} = 1 \pmod{n}.$$

This contradicts the second hypothesis and shows $k = 1$, i.e. $r = n - 1$. Hence $n - 1 = r \leq \phi(n) \leq n - 1$ so that $\phi(n) = n - 1$ and this shows that n is prime (compare 1.2).

3.5 Exercise   With the hints given below, prove the following slight improvement on 3.4: it says that we can choose different bases a for the different primes q. In fact if n *is* prime then a common base which works for all q will exist, using the existence of primitive roots − see Chapter 8, 2.3− but it may in practice be easier to find different bases for different primes q. In the examples we give below, it is always possible to find a common base fairly easily. Suppose $n > 1$ is odd and let the distinct primes dividing $n - 1$ be $q_1, q_2, \ldots, q_k$. Suppose that for each q_i there exists a_i such that

$$a_i^{n-1} \equiv 1 \ (\mathrm{mod} \ n) \ \text{and} \ a_i^{(n-1)/q_i} \not\equiv 1 \ (\mathrm{mod} \ n).$$

Prove that n is prime. [Hints. Let $d_i = \mathrm{ord}_n a_i$, so that $d_i \mid \phi(n)$ for each i. Let D be the least common multiple of the d_i, so that $D \mid \phi(n)$. Let α_i be the power to which q_i divides n. Use the hypotheses to show that $q_i^{\alpha_i} \mid d_i$ and deduce that $(n-1) \mid D$. Now deduce that $\phi(n) = n-1$.]

It is well worth amending program P4_3_4 (or P4_2_3 if you didn't incorporate Head's algorithm, but the larger examples below do require the greater precision of P4_3_4) as follows. We want to input n, then choose the base a and then input one by one the primes q (separately calculated) which divide $n - 1$. The main program of P4_3_4 is altered as follows (*newn* and q are new variables of type extended):

```
BEGIN    (* Main program *)
  writeln ('Type the value of n ');
  readln (nsaved);
  m:=nsaved;    (* We always work modulo the original n *)
  asaved:=2;    (* The first base we try is 2 *)
  writeln('a=2; input q=0 to go to another
      a and a=0 to finish');
  WHILE (asaved <> 0) DO
    BEGIN
      q:=1;    (* We first test n - 1 over q for q=1 *)
      WHILE (q <> 0) DO
        BEGIN
          n:=(nsaved - 1)/q;
          newn:=n;
          capT:=INT(SQRT(m)+0.5);
          t:=capT*capT-m;
```

```
        a:=asaved;
        PowerRule (a,n,m,r);
        writeln ('The value of (', asaved:0:0, ' to the
          power ', newn:0:0, ') mod ', m:0:0, ' is ',
          r:0:0);
        writeln('Type q');
        (* Enter q=0 to end this loop *)
        readln(q);
      END;
    writeln('Type a');
    (* Enter a=0 to end this loop *)
    readln(asaved);
  END;
END.   (* of main program *)
```

3.6 Computing exercises (a) Use Miller's test to show that $n = k \cdot 10^8 + 1$ is composite for $1 \le k \le 12$ except possibly for $k = 6$ and $k = 7$. Now use the $n-1$ over q test 3.4 to show that $k = 6$ and $k = 7$ do give primes. Remember that you may have to try several different bases before you find one which works. [Of course these are small enough to check by trial division too, given our agreement to allow this for numbers $< 10^{10}$.]

(b) Find all values of k, $1 \le k \le 70$, for which $k \cdot 10^{12} + 1$ is prime. (So start with Miller's test to eliminate most values of k.) You should find $k = 18$ is the first one, and base $a = 7$ works for that. A later value of k is $k = 63$, and it is slightly easier here to use 3.5 and find different bases for the different primes $q_i = 2, 3, 5, 7$.

(c) Show that $k \cdot 3^{25} + 1$ is composite for $1 \le k \le 33$ and that $k = 34$, 42 and 70 all give primes.

(d) Let $n = 10^{12} + 61$. Verify that $n - 1 = 2^2 \cdot 5 \cdot 3947 \cdot 12\,667\,849$ is the prime factorization of $n - 1$ and hence show that n is prime. Also show that $10^{12} + 63$ is prime by factorizing $10^{12} + 62$. Thus we have found a pair of *twin primes*.

(e) Here is an example where a two-stage process works. Let $n = 10^{13} + 37$, so that $n - 1 = 2^2 m$ where $m = 25 \cdot 10^{11} + 9$. Since $m > 10^{10}$, show it is prime by factorizing $m - 1$ and using 3.4. Then show n is prime by the same method.

(f) Verify that $10^{15} + 36 = 2^2 \cdot 7 \cdot 37 \cdot 965\,250\,965\,251$, and show that this last factor is prime by factorizing $965\,250\,965\,250$ and using the $n-1$ over q test. Hence show that $10^{15} + 37$ is prime.

(g) A fairly spectacular example in terms of the largeness of the base

a which needs to be taken to ensure $a^{(n-1)/2} \not\equiv 1 \pmod n$ is $n = 26\,437\,680\,473\,689$. You may like to verify that all numbers $a < 500$ give $a^{(n-1)/2} \equiv 1 \pmod n$. (Why is it is enough to verify this for primes a?)

Using the factorization $n - 1 = 2^3 \cdot 3^2 \cdot 11 \cdot 41^2 \cdot 89 \cdot 347 \cdot 643$, which you can easily verify using P2_1_13, prove that n is prime. This example will be referred to again in Chapter 11,1.6(b).

(h) Verify the factorization into primes $n - 1 = 2 \cdot 31 \cdot 258\,629\,069$ for $n = 16\,035\,002\,279$, and use the $n-1$ over q test to prove that n is prime. Verify also that $2n + 1$ is prime, for example by using 3.1. Do the same for $n = 16\,048\,973\,639$, factorizing $n - 1$ first.

(i) Show that $n = 30\,059\,924\,764\,123$ is prime using the $n - 1$ over q test (in contrast to (g) above the prime $q = 2$ gives no trouble here, but 3 gives a little trouble).

(j) In Chapter 5,2.2(b) you may have discovered the probable primes of the form $n = r^4 + 1$ for $1900 \le r \le 2000$. In fact, the values of r turn out to be $1900 + k$ where $k = 10$, 16, 26, 32, 34, 42, 44, 48, 52, 56, 62, 72, 78, 86, 94. (For extensive tables, see [Lal (1967)].) Finding the prime factors of $n - 1$ is here of course just a matter of finding the prime factors of r. Check that all these numbers n are in fact prime by the $n - 1$ over q test.

(k) Show that $1 + 8 \cdot 3 \cdot 5 \ldots 23 = 892\,371\,481$ is prime (the product contains all primes from 3 to 23).

(l) It can be shown that $10^{27} - 1 = 3^5 \cdot 37 \cdot 757 \cdot 333\,667 \cdot n$, where $n = 440\,334\,654\,777\,631$. Use program P2_1_15 to factorize $n - 1$ (mercifully it has only one prime factor $> 10^5$, and this is $< 10^{10}$) and hence show that the above is the prime factorization of $10^{27} - 1$. (This example occurs in [Lehmer (1927)].)

(m) This is a rather longer exercise. Using the identities $a^2 - b^2 = (a - b)(a + b)$ and $a^3 + b^3 = (a + b)(a^2 - ab + b^2)$, break down the number $10^{48} - 1$ as far as you can into factors. One of these factors is $10^{16} - 10^8 + 1$. Prove this is prime by the $n - 1$ over q test. Hence completely factorize $10^{48} - 1$ into primes, and hence completely factorize the 'repunit' $11 \ldots 1$ consisting of 48 ones in its decimal expansion.

The $n - 1$ over q test requires a knowledge of all the primes dividing $n - 1$; there is an improvement on this which requires only that 'most' of $n - 1$ is factorized.

3.7 Theorem *Suppose that $n - 1 = FR$ where all prime factors of F are known, and $(F, R) = 1$. Suppose that, for some a, we have*

$$a^{n-1} \equiv 1 \pmod n \text{ and } (a^{(n-1)/q}, n) = 1 \text{ for all primes } q \mid F.$$

Then each prime factor p of n satisfies $p \equiv 1 \pmod{F}$. In particular, if $F \geq \sqrt{n}$, then n is prime.

Proof Let q be a prime dividing F, and suppose that q^{α} is the exact power of q which divides F. If p is a prime dividing n, then we have, from the hypotheses of the theorem,

$$a^{n-1} \equiv 1 \pmod{p} \text{ and } a^{(n-1)/q} \not\equiv 1 \pmod{p}.$$

Let $r = \text{ord}_p\, a$. Then it follows that $r \mid (n-1)$ but $r \nmid (n-1)/q$. This implies that $q^{\alpha} \mid r$. [Write $n - 1 = q^{\alpha}t = rs$, where $q \nmid t$; then it is easy to see that $q \nmid s$, so that the whole power q^{α} must divide into r.] Since this is true for every prime $q \mid F$, we have $F \mid r$, and since $r \mid (p-1)$ by Fermat's theorem, we have $F \mid (p-1)$ as required. The last statement of the theorem just uses the fact that any composite number n has a prime factor $\leq \sqrt{n}$, and the smallest value of p is $F + 1 > \sqrt{n}$ here. $\qquad\square$

3.8 Corollary (Proth's theorem (F.Proth, 1878)).
Let $n = k \cdot 2^m + 1$ where $m \geq 2$, k is odd and $k < 2^m$. Suppose that there exists a such that $a^{(n-1)/2} \equiv -1 \pmod{n}$. Then n is prime.

Proof Let $F = 2^m$, $R = k$ in the theorem; then $k < 2^m$ implies $2^m \geq \sqrt{n}$ and the result follows. $\qquad\square$

The hypothesis $a^{(n-1)/2} \equiv -1 \pmod{n}$ is not unduly restrictive: if n is prime, then this power of a is $\pm 1 \pmod{n}$, and if it is $+1$ then it is worth trying another a. For an odd prime a, what one wants is that a is a 'quadratic nonresidue mod n'; compare Chapter 11,1.4(a) (where n and p stand for the a and n here) and also Chapter 11, 3.5(d).

3.9 Computing exercise Use Proth's theorem and a suitable program to determine the primes in the following series: $3 \cdot 2^m + 1$, $5 \cdot 2^m + 1$, $7 \cdot 2^m + 1$, all for $20 \leq m \leq 41$. You should check the compositeness of the numbers for which Proth's theorem fails, by using Miller's test. (For the first series the values of m are 30, 36 and 41.)

Here is another result that follows easily from 3.7:

3.10 Corollary *Suppose that $n = hp^k + 1$, where p is an odd prime and h is even, $h < p$ and $k \geq 1$. Suppose that there exists a such that $a^{n-1} \equiv 1 \pmod{n}$ and $(a^{(n-1)/p}, n) = 1$. Then n is prime.*

3.11 Project The result 3.10 can be used to find chains of primes rather like Cunningham chains (3.2 above). Taking $k = 1$ in 3.10, starting with $p = 3$ and $h = 2$ we have $2 \cdot 3 + 1 = 7$ is prime. Now take $h = 4$; we find $4 \cdot 7 + 1 = 29$ is prime. Write a program which finds small values of h with which to continue the series up to the limits of extended precision $(2 \cdot 29 + 1 = 59, \ 12 \cdot 59 + 1 = 709, \ldots)$.

4. Periods of decimals

One of the most delightful applications of the idea of order is to the study of the lengths of periods of decimals of the form $1/m$ (or indeed a/m where $(a, m) = 1$). For instance, $1/7 = 0.\overline{142857}$ (the bar indicating recurrence of this block of decimal places), while $1/11 = 0 \cdot \overline{09}$. Why does 7 have period length 6 and 11 have period length only 2? Gauss was sufficiently fascinated by this question that, as a schoolboy, he made a table of period lengths for m up to 1000. We shall assume that neither 2 nor 5 divides m; this has the effect (as we shall see) of making the period start immediately after the decimal point. Note that for example $1/15 = 0.0\overline{6}$, and $1/12 = 0.08\overline{3}$, the period starting one or two places to the right of the decimal point.

The ordinary calculation by division of the fraction $1/7$ is equivalent to the following sequence of divisions by 7:

$10 = 7 \cdot 1 + 3$ Now multiply the remainder 3 by 10 to get 30
$30 = 7 \cdot 4 + 2$ Now multiply the remainder 2 by 10 to get 20
$20 = 7 \cdot 2 + 6$ etc.
$60 = 7 \cdot 8 + 4$
$40 = 7 \cdot 5 + 5$
$50 = 7 \cdot 7 + 1$

The general step has the form $10r_j = 7q_j + r_{j+1}$, where $0 \le r_{j+1} < 7$. The decimal places of $1/7$ are the qs: $1/7 = 0.q_1q_2q_3 \ldots$. On the other hand it is the remainders which tell us when recurrence will happen: as soon as a 1 appears as a remainder r_j then the next line will start with 10 and the calculation will repeat. Indeed it is conceivable that a remainder other than 1 will occur which has occurred before, and recurrence will start from another place. We shall see that with $1/m$ for $(m, 10) = 1$ this does not happen.

For $1/m$ we get the following. Write $r_1 = 1$ by convention; then

$$10r_1 = mq_1 + r_2 \ (0 \le r_2 < m),$$

$$10r_2 = mq_2 + r_3 \ (0 \le r_3 < m),$$

and so on. *Assuming that $(m, 10) = 1$, note that the first line implies $(m, r_2) = 1$ and then the second line implies $(m, r_3) = 1$, and so on.* In particular none of the r_j can be zero (this would imply $m \mid r_{j-1}$) so in fact we have $0 < r_j < m$ for all j. Further it is clear that

$$r_2 \equiv 10 \ (\mathrm{mod} \ m), \ r_3 \equiv 10^2 \ (\mathrm{mod} \ m), \ldots, r_{j+1} \equiv 10^j \ (\mathrm{mod} \ m).$$

When j reaches $\mathrm{ord}_m 10$, we shall have $r_{j+1} = 1$ for the first time. We need to check that this is the first occasion on which any remainder equals a previous remainder. This will show that the calculation repeats from this point, and that the decimal for $1/m$ is $0.\overline{q_1 q_2 \ldots q_j}$ where $j = \mathrm{ord}_m 10$.

To check this, suppose that $r_{i+k} = r_i$ where $k \geq 1$ and i is as small as possible. We must prove that $i = 1$ (recall $r_1 = 1$). Suppose $i > 1$ and consider r_{i-1}: we have $r_i \equiv 10 r_{i-1}$ and $r_{i+k} \equiv 10 r_{i+k-1} \pmod{m}$, but $r_i = r_{i+k}$ and $(10, m) = 1$ so we get $r_{i-1} \equiv r_{i+k-1} \pmod{m}$ and hence these remainders must be equal: i was not the smallest after all. This contradiction shows that the calculation repeats from the beginning and not from some place after the beginning. We have proved:

4.1 Theorem *The length of the decimal period of $1/m$, where $(m, 10) = 1$, is the order of $10 \bmod m$, and the period starts from immediately after the decimal point.*

4.2 Computing exercises

(a) Write a program to determine, for a given m with $(m, 10) = 1$, the decimal expansion $0.q_1 q_2 q_3 \ldots$ of $1/m$, together with the remainders r_j. Stop when the period has been reached; your program automatically works out $\mathrm{ord}_m 10$ so print out this number.

(b) Use the program to find all the primes $p < 100$ for which $\mathrm{ord}_p 10 = p - 1$ (that is, 10 is a primitive root mod p; compare 2.7). Extend this to finding all $m < 100$ for which $\mathrm{ord}_p m = \phi(m)$, i.e. 10 is a primitive root mod m.

(c) A more ambitious version of (b), more like a project, is to find all $m < 1000$ for which 10 is a primitive root.

(d) Consulting 2.6(f), find all primes $p < 1000$ for which the length of the decimal period of $1/p$ is the same as that of $1/p^2$. Find the decimal expansions of these numbers. [You should find only two such primes.]

4.3 Project Test the conjecture: usually, the order of $10 \bmod p^a$ equals p^{a-1} times the order of $10 \bmod p$. (The case $a = 2$ is related to 2.6(f).)

4.4 Exercise: decimals a/m Let $m > 1$, $1 \leq a < m$, $(m, 10) = 1$ and $(a, m) = 1$. Suppose that the order of $10 \bmod m$ is the maximum possible, that is $\phi(m)$, so that 10 is a primitive root mod m. We work out the decimal expansion $0.Q_1 Q_2 Q_3 \ldots$ of a/m by the calculation

$$a = R_1,$$
$$10R_1 = mQ_1 + R_2 \quad (0 \leq R_2 < m),$$
$$10R_2 = mQ_2 + R_3 \quad (0 \leq R_3 < m),$$

and so on. Show that no R_i can be 0 and that $(R_i, m) = 1$ for all i. Why is it true that $a \equiv 10^b \pmod{m}$ for some b with $0 \le b < \phi(m)$? [Why in fact do the powers of 10 mod m cover *all* residues coprime to m?] Show that $R_i \equiv 10^{b+i-1} \pmod{m}$, and deduce that the period length of a/m is the same as that of $1/m$ (namely $\phi(m) = k$, say), while the remainders $R_1, R_2, R_3, \ldots, R_k$ are the cyclic permutation of the remainders r_1, r_2, $r_3, \ldots, r_k$ which has $R_1 = r_{b+1}$. Deduce that the digits $Q_1 Q_2 \ldots Q_k$ in the period are the same cyclic permutation of the digits $q_1 q_2 \ldots q_k$ in the period for $1/m$.

4.5 Project Write a program to find the decimal digits, remainders and period of a/m for any m coprime to 10 and any a with $(a, m) = 1$. For values of $m < 40$ say investigate the different sets of decimal digits which occur and relate the number of these different sets to $\phi(m)$ and the common length of all the periods.

4.6 Exercises

(a) Find the orders of 10 mod 7, 11, 13 and 17. Now use 4.1 and 2.6(i) above to find the lengths of the decimal periods of $1/m$ for $m = 7$, 11, 13, 17, 77, 91, 119, 143 and 221.

(b) Let p be a prime other than 2 or 5. Suppose that $1/p = 0.\overline{q_1 q_2 \ldots q_{2k}}$ has even period length $2k$. Show that $10^k \equiv -1 \pmod{p}$ and deduce that $r_{k+1} = p - 1$, using the same notation as above for the remainders. Show further that $r_2 + r_{k+2}$, $r_3 + r_{k+3}$, etc., are all equal to p. Show that, for each i with $1 \le i \le k$, we have $q_i + q_{i+k} = 9$. Use a program as in 4.2(a) to find values of p which give an even period and to verify this last result. (For example, $1/13 = 0 \cdot \overline{076923}$ and $0 + 9 = 7 + 2 = 6 + 3 = 9$.)

(c) Suppose that $(n, 10) = 1$. Show that, for all x and y,

$$10x \equiv 10y \pmod{2n} \Leftrightarrow x \equiv y \pmod{n}$$

and, if x and y are both even,

$$x \equiv y \pmod{n} \Leftrightarrow x \equiv y \pmod{2n}.$$

Show that, in the decimal expansion of $1/m$ where $m = 2n$, the remainders r_i for $i \ge 2$ are all even and the first repetition of remainders, $r_i = r_{i+j}$ (i minimal, $j \ge i$) occurs for $i = 2$. Deduce that the decimal expansion of $1/2n$ has the form $0.\overline{q_1.q_2 \ldots q_{k+1}}$ where k is the order of 10 modulo n. (Of course this is a special case of a theorem which generalizes 4.1 above by allowing powers of 2 and 5 to occur in the prime-power factorization of m. The general theorem is given, for example, in [K.M. Rosen (1988), p. 348].)

7

Cryptography

In this chapter we pause in the development of theoretical and computational tools to give a short account of one of the spectacular practical applications of number theory. This is to the sending of secure messages in secret, or encrypted, form. We shall concern ourselves mainly with one method of encryption, called RSA public key cryptography, the initials being those of the originators R.L. Rivest, A. Shamir and L.M. Adelman (1978). The basis of the security of this method is the tremendous difficulty of factoring very large numbers. The method also requires the generation of very large primes (100 digits or so) and in practice these are found by probabilistic methods as in Section 2 of Chapter 5. Of course we are restricted here to rather unrealistically small primes, in order to stay within extended precision, but, as with the other subjects treated in this book, a very adequate idea can be obtained from these modest examples.

There are many existing accounts of public key cryptography; for example Chapter 7 of [K.H. Rosen (1988)], Chapter IV of [Koblitz (1987)], Chapter 6 of [Riesel (1985)] and [Salomaa (1990)], not to mention more popular accounts such as Chapters 13 and 14 of [Gardner (1989)] and [Hellman (1979)]. We shall concentrate here on relating the basic ideas to the theory which has been developed so far, and will give a program which illustrates these ideas. Various suggestions will be made for developing this program. On the whole the additional programming needed is not of a particularly 'mathematical' kind, involving rather the manipulation of strings of characters, and readers who do not care for this style of programming could regard this chapter as optional reading. (We shall not, in any case, develop essential new theory here.) It may, all the same, be of interest to read the initial theory to see what it is all about.

1. Exponentiation ciphers

These, like their more sophisticated relatives, the RSA ciphers discussed below, date from 1978. They were invented by S. Pohlig and M. Hellman. The idea is simple enough. We choose a prime p and an 'enciphering key' e, which is coprime to $\phi(p) = p - 1$. The 'plaintext' message which is to be enciphered is first converted into an entirely numerical form (one method for doing this is described below) and broken up into blocks of digits, all of which, when regarded as ordinary numbers, are $< p$.

Suppose for example that $p = 37\,781$ and that the blocks of digits in the numerical form of the message are

$$8069\ 6578\ 8584\ 8332$$

To encrypt this, we replace each four-digit number P (standing for 'Plaintext' though of course the four-digit numbers are one step removed from what most people regard as plain English) by $C \equiv P^e \pmod{p}$, $0 < C < p$. Thus choosing $e = 367$ (which is coprime to $p - 1 = 37\,780 = 2^2 \cdot 5 \cdot 1889$) the encrypted form is (using one of the power rule programs P4_2_3 or P4_3_4)

$$37059\ 12498\ 22955\ 20254$$

(that is, $8069^{367} \equiv 37\,059 \pmod{p}$, etc.).

How can this message be 'decrypted', that is returned to the first numerical form and thence back to ordinary English? Let d (the deciphering key) be such that $de \equiv 1 \pmod{p - 1}$. This is just the *inverse* of e modulo $p - 1$, which will exist since $(e, p - 1) = 1$, and which can be calculated by the method of Chapter 1, 3.17(a). In the present case $d = 3603$. Now

$$C^d = (P^e)^d = P^{ed} = P^{ed-1}P \equiv P \pmod{p}$$

by Fermat's theorem, since $(p - 1) \mid (ed - 1)$. Thus raising each of the encrypted (five-digit) numbers to the power d and reducing mod p, we recover the original (four-digit) numbers. Thus $37\,059^{3603} \equiv 8069 \pmod{p}$, etc.

In order to use this for sending secret messages between two people A and B, both will need to know the prime p (all encryptions are done modulo p), and A will need to choose an enciphering key e, while B will need to work out the corresponding deciphering key d. Of course B

could choose a different enciphering key e' for sending messages to A, so long as A knows the corresponding deciphering key d' with $d'e' \equiv 1$ (mod $p - 1$).

Breaking such a cipher presents problems, even if p is known, and if one corresponding plaintext P and ciphertext C are known, so that $C \equiv P^e$ (mod p). The problem is that it is not an easy matter to determine e from such a congruence when p is a large prime. (A description of this problem, and some solutions, are in [Koblitz (1987), pp.98ff.].)

There are of course many methods for turning a message consisting of letters (and spaces) into a sequence of numbers. The simplest may be to associate A with 01, B with 02, ..., Z with 26, and the space with 00. But we are dealing here with computers, and they have an in-built coding for letters (and all other symbols), namely the ASCII code. There are advantages in using this code since it removes the necessity of setting up a special letter–number correspondence in any programs we write.

The ASCII codes for the capital letters $A, B, \ldots, Z$ are $65, 66, \ldots, 90$ respectively, and the space has code 32. PASCAL has functions available to transfer from characters to integers and vice versa; thus

$$\mathrm{ORD}(A) = 65, \mathrm{ORD}(B) = 66, \ldots, \mathrm{ORD}(Z) = 90, \mathrm{ORD}() = 32,$$

$$\mathrm{CHR}(65) = \text{`}A\text{'}, \mathrm{CHR}(66) = \text{`}B\text{'}, \ldots, \mathrm{CHR}(90) = \text{`}Z\text{'}, \mathrm{CHR}(32) = \text{` '}.$$

In the program given below (P7_2_3) we use this correspondence. (It was also used to produce the example above; what in fact is the English equivalent of the sequence of 4-digit numbers?) The ASCII codes for lower-case letters $a, b, \ldots, z$ are $97, 98, \ldots, 122$ respectively so to avoid three-digit codes we shall stick to capital letters. (We also, to simplify matters, ignore punctuation.)

Having turned each individual character into a (two-digit) number there are several ways to produce numbers corresponding to several consecutive characters in a message. We shall stick here to the most straightforward method of juxtaposition: if G is 71 and P is 80 then GP is 7180. (Using the numbers $00, 01, \ldots, 26$ to represent space, $A, \ldots, Z$ an alternative would be to use numbers to base 27: C GAUSS would become $3 \cdot 27^6 + 0 \cdot 27^5 + 7 \cdot 27^4 + 1 \cdot 27^3 + 21 \cdot 27^2 + 19 \cdot 27 + 19 = 1\,166\,017\,078$ instead of $03\,000\,701\,211\,919$ which it becomes through juxtaposition.)

With an exponentiation cipher as above, we choose p and then break the message up into blocks of characters, the number of characters, c, in a block being as large as possible consistent with

$$90 \ldots 90 < p$$

where there are c 90s on the left. (This is because the largest ASCII code being used is 90.) Thus with $p = 37\,781$ as above, we can take $c = 2$ since $9090 < 37\,781$. The message will be broken into blocks of length 2 characters (each one a capital letter or a space), and these converted into numbers with four digits ready for encryption.

In order to write a program to implement exponentiation ciphers (or the RSA ciphers of Section 2 below) it is necessary to incorporate the power algorithm from Chapter 4. In $P7\_2\_3$ below we give a program which implements RSA ciphers; these involve two primes, p and q, and the program is arranged so that by putting $q = 1$ it also implements an exponentiation cipher with prime p. We hope that this program and its commentary show the working of encryption and decryption sufficiently clearly for you to adapt it to your own purposes.

2. Public key cryptography: RSA ciphers

If the prime p and the enciphering key (exponent) e are both discovered in an exponentiation cipher then of course all is lost (or found): the deciphering key can easily be worked out as the inverse of e modulo $p-1$. Public key cryptography is an ingenious method of sending secure messages between two (or any other number of) people, even when all the enciphering keys are made public (hence the name), because it is completely infeasible to find the deciphering keys.

The idea is, again, very simple. If Rosencrantz and Guildenstern (referred to as R and G (respectively)) wish to set up secure communication, each chooses two large primes: p_R, q_R and p_G, q_G say. Let $n_R = p_R q_R$, $n_G = p_G q_G$. The numbers n_R and n_G can safely by made public, provided the primes used are all large (compare Chapter 2, 3.5). Each chooses an enciphering key: e_R and e_G say, coprime to $\phi(n_R)$, $\phi(n_G)$ respectively. (Note that $\phi(pq) = (p-1)(q-1)$ when p and q are prime.) These enciphering keys can, too, be made public. Each person, knowing his p, q and e, can easily compute an inverse d for e (mod $\phi(n)$). But an outsider, knowing only n and e, has little chance of computing d without actually factorizing n (compare 2.1(b) below), and indeed R is unable to calculate d_G and G is unable to calculate d_R.

When Rosencrantz wants to send a message to Guildenstern he converts it in the usual way into numerical blocks. He then uses *Guildenstern's* enciphering key e_G (which, like all enciphering keys, is public knowledge) to encipher each block P according to the *enciphering rule* E_G:

$$E_G(P) \equiv P^{e_G} \pmod{n_G}.$$

Note that having enciphered it, R would be unable to decipher it again, since he does not know d_G. However G is not under the same disadvantage: knowing both d_G and n_G he can decipher by means of the *deciphering rule*

$$D_G(C) \equiv C^{d_G} \pmod{n_G}.$$

Thus $D_G(E_G(P)) \equiv P \pmod{n_G}$, by the same argument as was used in the exponentiation cipher case in Section 1.

Of course, when G wishes to send a message to R, he uses the enciphering rule

$$E_R(P) \equiv P^{e_R} \pmod{n_R},$$

and R deciphers with the deciphering rule

$$D_R(C) \equiv C^{e_R} \pmod{n_R}.$$

Again, $D_R(E_R(P)) \equiv P \pmod{n_R}$.

Remember that R knows E_R, E_G and D_R, while G knows E_R, E_G and D_G.

An important part of a written message is always the *signature*, whose individuality convinces the recipient that the message really did come from the person claiming to be the sender. Remarkably, it is possible to send signatures using the RSA public key system, too. Let us suppose that $n_R < n_G$ (of course both R and G know this) and that R wishes to send a signature to G, that is a sequence of characters at the end of his message which G can make sense of by deciphering and which could only have come from R. Suppose in fact that the signature is one block P of plaintext. Then R enciphers it by working out $D_R(P)$ and then $S = E_G(D_R(P))$. Since $n_R < n_G$ there will be no need to split $D_R(P)$ into smaller blocks before further enciphering it as S. At the end of his message, R then sends the signature S. When the recipient G deciphers the message using the deciphering key d_G, there will be some gibberish, namely

$$D_G(S) = D_G(E_G(D_R(P))) = D_R(P)$$

at the end. (Of course $D_R(P)$ itself is a number. The gibberish results when G tries to turn this into words.) Presuming this to be R's signature, he applies E_R to this, which recovers P; when turned into words this will now make sense. Note that only the sender R knows D_R so only he could have sent the particular gibberish $D_R(P)$ at the end of his message.

2.1 Exercises (computing and otherwise)

(a) If n is the product of two odd primes, $n = pq$, and if p and q are nearly equal, then there is a method due to Fermat for factorizing n. It works as follows. Note that $n = pq = \left(\frac{p+q}{2}\right)^2 - \left(\frac{p-q}{2}\right)^2$. Write this as $t^2 - s^2$ say, where s will be *small*, and so t will not be much larger than $\sqrt{n}$. Starting with $t = [\sqrt{n}] + 1$ we work out $t^2 - n$ and check to see if it is a square; if not then add 1 to t and repeat. When $t^2 - n = s^2$ then of course $n = (t + s)(t - s)$.

The moral of this from the point of view of public key cryptography is that the two primes p and q should not be chosen too close; in practice a difference of three or four in the number of digits is quite enough to make Fermat's method impracticable. Write a program to implement Fermat's method. Try $n = pq$ where $p = 57\,685\,933$ and $q = 57\,700\,183$; try the same p and $q = 58\,000\,171$. Note the difference in the number of steps taken in these two cases! You can manufacture your own examples by using program P2_2_1 to find primes close to chosen integers. You may be surprised at how often the method terminates after one step (i.e. with $t = [\sqrt{n}] + 1$) when the factors p and q are quite close. (If you tried 4.5(a) of Chapter 1, then you will already have met $t^2 - n$ for this t.)

Try factorizing $n = 1\,112\,470\,797\,641\,561\,909$ (which happens to be $10^{22} + 1$, divided by the relatively small factors 89 and 101) using your program.

(b) Here is an argument to find, for a given prime p, how large the prime $q > p$ can be while Fermat's method still takes only one step to factorize $n = pq$ (see (a) above). The value of t begins at $t = [\sqrt{n}] + 1$ and, when $t = (p + q)/2$, the method terminates and factors are found. Thus the number of steps needed is

$$\frac{p+q}{2} - [\sqrt{n}] = \frac{1}{2}(p+q) - [\sqrt{pq}].$$

This is $< \frac{1}{2}(p+q) - \sqrt{pq} + 1$, which is ≤ 2 if and only if

$$q^2 - 2q(p+2) + (p-2)^2 \leq 0.$$

Check that this holds (given $q > p$) if and only if $q \leq p + 2 + 2\sqrt{2p}$. Consequently, if this holds, then the number of steps is certainly < 2. On the other hand, the number of steps is $> \frac{1}{2}(p+q) - \sqrt{pq}$ (since pq is not a perfect square), and writing down the condition for this to be ≥ 1(so that the number of steps is certainly > 1) shows that the above estimate for q is exact: if $q \geq p + 2 + 2\sqrt{2p}$ then at least two steps are needed. (Of course we can never have equality here for $p > 2$.)

For example with $p = 11\,069$ this gives $q \leq 11\,368$, and the largest prime $\leq 11\,368$ is $11\,353$. In fact the next prime, $11\,369$, does indeed need two steps of the method.

Try some other primes p and check that the estimate gives the largest prime for which Fermat's method terminates after one step. Note that if p and q are odd but not both primes then Fermat's method may terminate more quickly than the above argument suggests. For the method may discover a factor *between* $\sqrt{pq}$ and q. As an example, let $p = 3$, $q = 15$: the method discovers the factor 9 first (after one step), whereas for *primes* q the above argument shows that two steps are needed for $q \geq 11$.

Can you estimate the largest q for which *two* steps of the method will suffice, assuming p and q are prime?

(c) From a knowledge of n and $\phi(n)$ (and knowing $n = pq$ for distinct primes p and q) it is easy to find p and q. For $\phi(n) = (p-1)(q-1)$ and we have $pq = n$, $p+q = n+1-\phi(n)$. Hence p and q are the roots of the quadratic equation $x^2 - (n + 1 - \phi(n))x + n = 0$. Write a program to determine p and q from n and $\phi(n)$. Test with small numbers and then with n around 10^{16}, finding primes p and q to multiply together using P2_2_1.

(d) There is a slight possibility that the plaintext number P may not be coprime to the product $pq = n$ (the probability of this is $(n-\phi(n))/n$, that is $1/p+1/q-1/pq$, and this is very small when p and q are reasonably large). Show that nevertheless, if $de \equiv 1 \pmod{\phi(n)}$, then $P^{de} \equiv P \pmod{n}$. [This is quite similar to Chapter 4, 4.10(c) and depends on the fact that n is a product of *distinct* primes. In fact it holds for n equal to any product of distinct primes.] Of course, if $(P, n) > 1$ then the gcd of P and n must be p or q, since by choice $P < n$. Thus a plaintext message *not* coprime to n is disastrous for keeping the factors p and q secret!

(e) Suppose $n_G < n_R$. Show that R can send his signature to G by reversing the order of the operations E_G and D_R above. Explain how this can be deciphered and why it is a guaranteed signature.

2.2 Project   Write a program which will simulate the sending of messages between two people. All the ingredients necessary are present in the program P7_2_3 below. Incorporate the signature sending facility described above.

2.3 A sample program for public key cryptography   The program below incorporates the power algorithm, encryption and decryption according to the procedure given above for RSA ciphers (or, putting $q = 1$, for exponentiation ciphers). In particular there is a procedure for finding the inverse of the encryption exponent e modulo $\phi(n)$, which is essentially P1_3_16. The program uses the ACSII code for transcribing messages consisting of (capital) letters and spaces into numbers. Since

the power algorithm uses Head's algorithm for the multiplication on
numbers modulo n, it should give accurate results for n up to about
10^{18}.

As it stands, the program prompts you for two primes p and q, then
prints out $n = pq$, then prompts you for an encryption key e (called
enc in the program). If this is not coprime to $\phi(n)$ then you will be
re-prompted for e, and eventually the inverse d ($= dec$) of a suitable e
will be found. You are then prompted for the length of block; remember
that if ℓ is the length then $9090\ldots90$ with ℓ occurrences of 90 must be a
number $< n$, in order that the numerical equivalent of every combination
of letters and spaces in a block will be $< n$. You are then prompted for
the number of blocks and invited to type your message. The program
then encrypts and immediately decrypts it, so you can see whether it is
working!

Here is an example.   Take $p = 1\,546\,379$, $q = 365\,738\,333$, then
$n = 565\,570\,077\,646\,207$. Take $e = 12\,345$; then $d = 208\,864\,504\,620\,337$.
With block length 7, message length 21 and message 'MAFEKING IS
RELIEVED ' (with one space at the end), the numerical form of the
message is

    77657069757378   71327383328269   76736986696832

and the encrypted form is

    359957508050520   165550807112978   467553571638325.

```
PROGRAM P7_2_3;

VAR
message_length,num_blocks,block_length,i,j :  integer;
s,numstring :  string;
x:  array[1..32] of char;
a,b,c:  array[1..32] of extended;
power_of_ten, max_power:  extended;
t, capT: extended;
n,p,q,phi:extended;
(* n is the product of two primes p and q *)
enc,dec:extended;    (* enc is the encrypting exponent, *)
(* and must be coprime to (p-1)(q-1).  *)
(* dec is the decrypting exponent.  *)
u:integer;
success:boolean;    (* This is used to signal when an *)
(* encrypting exponent has been found which is coprime *)
(* to (p-1)(q-1) *)
```

```
PROCEDURE MultiplyModN (x,y,modulus:  extended;
      VAR prod:  extended);
(* Multiplies x and y modulo 'modulus' (here this *)
(* is n) using Head's algorithm and stores the *)
(* result in prod *)

VAR a, b, c, d, z, e, f, v, v1, g, h, j, j1, k:
extended;

  BEGIN

  a:=INT(x/capT); b:=x - a* capT;
  c:=INT(y/capT); d:=y - c* capT;
  z:=a*d + b*c;
  z:=z - modulus*INT(z/modulus);
  e:=INT(a*c/capT); f:=a*c-e*capT;
  v:=z + e*t;
  v1:=v/modulus;
    IF (v>0) OR (v1=INT(v1)) THEN v:=v-modulus*INT(v1)
    ELSE v:=v-modulus * (INT(v1)-1);
  g:=INT(v/capT); h:=v - g*capT;
  j:=(f + g)*t;
  j1:=j/modulus;
    IF (j>0) OR (j1=INT(j1)) THEN j:=j-modulus*INT(j1)
    ELSE j:= j-modulus*(INT(j1)-1);
  k:=j + b*d;
  k:=k-modulus*INT(k/modulus);
  prod:=h*capT+k;
  prod:=prod-modulus*INT(prod/modulus);

END;   (* of MultiplyModN *)


PROCEDURE PowerRule (base, power, modulus:extended;
      VAR res:  extended);
VAR e:  extended;
BEGIN
  res:=1;
  WHILE power>0 DO
    BEGIN
      e:=power-2*INT(power/2);
      IF e=1 THEN
        MultiplyModN(base,res,modulus,res);
      MultiplyModN(base,base,modulus,base);
```

```
      power:=(power-e)/2;
    END;

END;    (* Of PowerRule *)


PROCEDURE Inverse(a:  extended; VAR inv:  extended);
(* This finds the inverse of a modulo phi *)
(* (=(p-1)(q-1), or p-1 when q=1) when they are *)
(* coprime, and stores the result in inv *)

VAR b,a0,b0,r,s,s1,s2,t,t1,t2,q,quotient:  extended;

BEGIN
  b:= phi; a0:=a; b0:=b;
  s2:=1; s1:=0;
  t2:=0; t1:=1;
  WHILE b > 0 DO
    BEGIN
      quotient:=a/b;
      IF ((quotient>0) OR (quotient = INT(quotient)))
        THEN q:=INT(quotient)
      ELSE
        q:=INT(quotient)-1;
      r:=a - b * q;
      s:=s2 - q * s1;
      t:=t2 - q * t1;
      a:=b;
      b:=r;
      s2:=s1;
      t2:=t1;
      s1:=s;
      t1:=t;
    END;
  IF (a=1) THEN
    BEGIN
      writeln(a0:0:0,' and ',b0:0:0, ' are coprime');
      quotient:=s2/b0;
      IF ((quotient>0) OR (quotient = INT(quotient)))
        THEN s2:=s2-INT(quotient)*b0
      ELSE
        s2:=s2-(INT(quotient)-1)*b0;
      inv:=s2;
```

```
        success:=true;
      END
    ELSE
      writeln('Your choice of e is not coprime to
      (p-1)(q-1)');
  END;   (* of Inverse *)


BEGIN   (* main program *)
  writeln('Type primes and q (or p prime and q = 1) ');
  readln(p,q);
  n:=p*q;

  capT:=INT(SQRT(n)+0.5);
  t:=capT*capT-n;   (* These are needed in MultiplyModN *)

  IF (q>1) THEN   (* Using just a single prime *)
    phi:=(p-1)*(q-1)   (* as in exponentiation ciphers *)
  ELSE   (* we can put q=1 *)
    phi:=p-1;

  writeln('n = ',n:0:0);
  success:=false;
  WHILE (success=false) DO
    BEGIN
      writeln('Type e, the encryption exponent');
      readln(enc);
      Inverse(enc,dec);
    END;

  writeln('Inverse is ',dec:0:0);

    writeln('Enter the block length');
  (* This refers to the number of CHARACTERS *)
  (* in a block.  Using just the letters A..Z and the *)
  (* space, all ASCII codes are <=90.  *)
  (* Hence 9090..90 with block_length 90s must be < n *)
  readln(block_length);

  writeln('Enter the message length');
  readln(message_length);

  writeln('Type the string in CAPITALS');
  readln(s);

  num_blocks:=message_length div block_length;
```

```
(* Integer division !*)
FOR i:= 1 TO num_blocks DO
  BEGIN
    a[i]:=0;
    FOR j:=1 TO block_length DO
      BEGIN
        x[j]:=s[j+block_length*(i-1)];
        a[i]:=a[i]*100 + ord(x[j]);
      END;
    writeln('This block becomes the number ',a[i]:0:0);
    (* This is the (2*block_length)-digit number *)
    (* associated with the i-th block of characters *)

    Powerrule(a[i],enc,n,b[i]);
    writeln('Encrypted equivalent of this block
        is ',b[i]:0:0);
    (* b[i] is the encrypted equivalent of *)
    (* the i-th block of characters *)

    Powerrule(b[i],dec,n,c[i]);
    writeln('Applying the inverse gives ',c[i]:0:0);
    (* This should give the same number as a[i] !*)
    writeln;

  END;
(* Now we need to turn the numbers c[i] back into *)
(* characters so we need to read the number c[i] in *)
(* groups of two digits, from left to right *)
writeln('The decrypted message is:  ');
writeln;

power_of_ten:=1;
FOR j:=2 TO block_length DO
  power_of_ten:=power_of_ten*100;
max_power:=power_of_ten;

FOR i:=1 TO num_blocks DO
  BEGIN
    power_of_ten:=max_power;
    FOR j:=1 TO block_length DO
      BEGIN
        u:=TRUNC(c[i]/power_of_ten);
        c[i]:=c[i] - u*power_of_ten;
```

```
            power_of_ten:=power_of_ten/100;
            write(chr(u));
         END;
      END;
   writeln;
END.
```

3. Coin-tossing by telephone

We conclude this chapter with another application of the difficulty of factoring large numbers. If a random decision has to be made between two alternatives (such as who should play white in a game of chess, or who should serve first in a game of tennis), then a standard method of making the decision is to toss a coin (or a racquet). The idea is that one person guesses the outcome of the random event, and, if the guess is correct, that person then 'wins the toss' and has the privilege of choosing what will happen (who will play white, etc.). It is important that both people involved agree as to whether the guess was correct! If you imagine doing this over the telephone, rather than in the physical presence of both parties, then the difficulties become obvious. How can one person who sees the coin come down tails convince the other person, who guessed heads, that the guess was wrong?

We shall present the method in the form of a sequence of exercises, since there are a few mathematical details to fill in.

3.1 Exercises

(a) Let p be a prime of the form $4k + 3$ and let a and b satisfy $a \equiv b^2$ (mod p), where $p \nmid a$ (equivalently $p \nmid b$). Fermat's theorem shows that $b^{4k+2} \equiv 1$ (mod p). Use this to show that $(a^{k+1})^2 \equiv a$ (mod p), i.e. that $x = a^{k+1}$ is a solution to $x^2 \equiv a$ (mod p). (The role of b here is that a has to be congruent to a square, modulo p. We shall meet this result again in Chapter 8, 2.4(f).) Note that $x^2 \equiv a$ (mod p) $\Leftrightarrow x^2 \equiv b^2$ (mod p) $\Leftrightarrow x \equiv \pm b$ (mod p), since p is prime, so $b = \pm a^{k+1}$ and there are exactly two solutions, $x = \pm a^{k+1}$, to the congruence $x^2 \equiv a$ (mod p).

(b) Let $n = pq$ where p and q are distinct primes, both congruent to 3 (mod 4), and let $a \equiv b^2$ (mod n) as above. Using the fact that

$$x^2 \equiv a \text{ (mod } n) \Leftrightarrow x^2 \equiv a \text{ (mod } p) \text{ and } x^2 \equiv a \text{ (mod } q),$$

deduce that there are exactly *four* solutions to $x^2 \equiv a$ (mod n).

To find them explicitly, let r and s be such that $rp + sq = 1$ (recall p and q are distinct primes, so $(p, q) = 1$), and let $x_1 \equiv a$ (mod p), $x_2 \equiv a$

(mod q). (The method of (a) above can be used to find x_1 and x_2.) Show that

$$x = x_1 sq + x_2 rp$$

satisfies $x^2 \equiv a$ (mod p and mod q) and hence (mod n). By filling in the two choices for x_1 and the two choices for x_2 this gives four solutions (why are they necessarily distinct (mod n)?).

(c) Now we come to the coin-toss over the telephone. Our two protagonists are again R and G.

(i) R chooses two large primes p and q, both congruent to 3 modulo 4, and he sends G the product $n = pq$.

(ii) G chooses at random a number b with $1 \leq b \leq n-1$, and computes $a \equiv b^2$ (mod n). (Note that $a \not\equiv 0$ (mod n); why is this?) G returns a to R.

(iii) R determines the four solutions to $x^2 \equiv a$ (mod n), using the method of (a) and (b) above. Remember that R knows the factorization of n. These solutions are $\pm b$ (where b is as in (ii)) and two others, say $\pm c$.

(iv) Now comes the toss. R sends back to G one of the four solutions he has found. The essential difference is between $\pm b$ or $\pm c$. The agreement is that if R sends back $\pm b$ to G then R has 'won' the toss; if R sends back $\pm c$ to G then R has 'lost' the toss. So R 'guesses' which solution G started with, and if he guesses right then he wins. Since there is no way of telling which solution G did start with the guessing is fair.

(v) Now comes the verification of a win or a loss. Suppose that G receives the solution $\pm c$ from R, so that R has lost the toss. G proves this as follows. Adding together two solutions to get $b + c$, G works out $\gcd(b + c, n)$. Show that this is either p or q (using the fact that b and c differ just in the sign of x_1 or x_2 in the formula for x in (b) above). Hence G can factor n ! There is no reasonable way he could do this, provided p and q are large, without the information gained from the two solutions. (The only possible remaining objection is that G presumably cannot convince R that R has won!)

3.2 Project Write a program to implement this method of tossing coins by telephone.

8

Primitive roots

In Chapter 6,2.7, we defined a *primitive root* mod n to be a number g whose order mod n is $\phi(n)$. Such a g is necessarily coprime to n. In Section 2 below we turn to the question of which n possess primitive roots but first let us see what possible use it can be to know that g is a primitive root mod n. Some of the results below were first written down before the time of Gauss, e.g. by J.H. Lambert (1769) and L. Euler (1773), but the first fully correct treatment of the subject occurs in C.F. Gauss's *Disquisitiones Arithmeticae* (sections 52–55), published in 1801. (See [Gauss (1986)].)

1. Properties and applications of primitive roots

1.1 Proposition   *Suppose that g is a primitive root* mod n. *Then*
   (a) *for any integers r and s, $g^r \equiv g^s \pmod{n} \Leftrightarrow r \equiv s \pmod{\phi(n)}$,*
   (b) *the $\phi(n)$ numbers $1, g, g^2, \ldots, g^{\phi(n)-1}$ are all distinct* mod n. *These are therefore,* mod n, *precisely the numbers coprime to n ; in particular their least positive residues are precisely the positive integers $\leq n$ and coprime to n.*

Proof
   (a$\Rightarrow$) We may suppose that $r \geq s$. Then $g^{r-s}g^s \equiv g^s \pmod{n}$, and, since $(g, n) = 1$, we can cancel g^s from both sides to give $g^{r-s} \equiv 1 \pmod{n}$. Thus (see Chapter 6, 2.5), $\phi(n) = \mathrm{ord}_n g \mid (r - s)$, which gives the result. The reverse argument is valid and proves (a$\Leftarrow$). (b) This is really

an application of (a), since no two of the powers $0, 1, 2, \ldots, \phi(n) - 1$ can be congruent mod $\phi(n)$.   □

(In group-theoretic terms, (b) just says that a primitive root is a generator of the cyclic group consisting of residues mod n which are coprime to n.) The above result is useful for solving congruences involving exponents, and we illustrate the method by means of examples. In fact we have already met one similar example, in Chapter 6,2.12.

1.2 Examples

(a) In Chapter 4,3.5(b) some solutions to the congruence $a^{p-1} \equiv 1$ (mod p^2) were found, a being given and p being prime. If, instead, p is given, and if we can find a primitive root g mod p^2 (something that always exists, as we shall see below), then it is relatively easy to find all the values of a which satisfy the congruence. Since (a, p^2) must be 1 for the congruence to have any solutions at all, we are assured that $a \equiv g^x$ (mod p^2) for some value of x, by 1.1(b).

We have $a^{p-1} \equiv g^{(p-1)x}$ (mod p^2), which is $\equiv 1$ (mod p^2) if and only if $(p-1)x \equiv 0$ (mod $\phi(p^2) = p(p-1)$). This holds if and only if $x \equiv 0$ (mod p), so the solutions are $a \equiv g^x$ (mod p^2) where $x = 0, p, 2p, \ldots, (p-2)p$. For instance, when $p = 7$ it is easy to verify that 3 is a primitive root (mod 49, and also mod 7 in fact), and these solutions come to $a \equiv 1, 18, 19, 30, 31, 48$ (mod 49). Of course, $p - 1$ being even, if a is a solution then so is $-a$.

Note that there are always exactly $p-1$ distinct solutions (mod p^2). It is in fact true that $a^{p-1} \equiv 1$ (mod p^k) has exactly $p-1$ distinct solutions (mod p^k) for any $k \geq 1$. This should be compared with the difficulty of finding p with $a^{p-1} \equiv 1$ (mod p^2) for a *given* a. (Chapter 4,3.5(b).)

(b) Let us solve the equations $12^x \equiv 17$ (mod 25) and $y^9 \equiv 17$ (mod 25). In fact, 2 is a primitive root mod 25, as can be seen from the following table showing powers of 2 mod 25. Recall that $\phi(25) = 20$.

k	0	1	2	3	4	5	6	7	8	9	10	11	12	13	14	15	16	17	18	19
2^k	1	2	4	8	16	7	14	3	6	12	24	23	21	17	9		18	11	22	19	13

Not only does this show that 2 is primitive, since $2^k \not\equiv 1$ (mod 25) for $0 < k < 20$, but it gives the solution k to $2^k \equiv b$ for each b coprime to 25. In this situation we sometimes say that b has *index* k (mod 25), the primitive root 2 being understood. Indices have a close relationship to logarithms since multiplying numbers adds their indices.

Let us return to $12^x \equiv 17$ (mod 25). Since, mod 25, $12 \equiv 2^9$ and $17 \equiv 2^{13}$ from the table, we can rewrite this as $2^{9x} \equiv 2^{13}$ (mod 25), and, from 1.1(a), this is equivalent to $9x \equiv 13 \pmod{20}$. Solving this in the usual way (e.g. $9 \times 9 \equiv 1 \pmod{20}$) gives $x \equiv 17 \pmod{20}$.

As for $y^9 \equiv 17 \pmod{25}$, we note that this implies $(y, 25) = 1$, so we can write $y \equiv 2^x \pmod{25}$ and the equation reads $2^{9x} \equiv 2^{13} \pmod{25}$ which, as before, gives $x \equiv 17 \pmod{20}$. Thus, from the table again, $y \equiv 22 \pmod{25}$.

1.3 Exercises

(a) Verify that 2 is a primitive root mod 11^2 and mod 13^2. (Possibly it is worth using a short program to calculate the order, simply multiplying 2 by itself and reducing mod p^2 each time until 1 is reached.) As in 1.2(a) above, verify that the solutions to $a^{p-1} \equiv 1 \pmod{p^2}$ are, for $p = 11$, $a \equiv 1, 3, 9, 27, 40, 81, 94, 112, 118, 120 \pmod{121}$, Use Program P4_2_3 to calculate the powers of 2. Find the solutions for $p = 13$.

(b) Verify that 7 is a primitive root mod 22. Hence find all solutions x to $19^x \equiv 17 \pmod{22}$.

(c) Verify that 3 is a primitive root mod 17 and hence show that $7^x \equiv 6 \pmod{17} \Leftrightarrow x \equiv 13 \pmod{16}$.

(d) Verify that 5 is a primitive root mod 18 and hence show that $13^x \equiv 7 \pmod{18} \Leftrightarrow x \equiv 2 \pmod{3}$.

(e) Suppose that g is a primitive root mod n, where $n > 2$. Show that $x^2 \equiv 1 \pmod{n}$ has exactly two solutions (write $x \equiv g^k$) and deduce that

$$x^2 \equiv 1 \pmod{n} \Leftrightarrow x \equiv \pm 1 \pmod{n}.$$

2. Existence and nonexistence of primitive roots

In 2.3(d) of Chapter 6 it is shown (modulo details) that, if n is a product of two coprime integers, both > 2, then n has no primitive root. We hope a little thought convinces you that this establishes the following result.

2.1 Lemma *The only possible numbers n which can possess primitive roots have the form $n = 2^k$ or p^m or $2p^m$ where $k \geq 0$, $m > 0$ and p is an odd prime.*

We have also seen, in Chapter 6,2.6(c), that 8 has no primitive root, so not all of the above possess primitive roots. In fact we can rule out 2^k for $k \geq 3$, as follows. (See also 2.4(d) below.) Note that 2 and 4 *do* have primitive roots, namely 1 and 3 respectively.

2.2 Proposition *Suppose that a is a primitive root mod 2^k for some*

$k \geq 3$. *Then a is also a primitive root* mod 2^{k-1}. *Hence there are no primitive roots* mod 2^k *for* $k \geq 3$.

Proof Clearly, a is odd. We have $a^{2^{k-2}} \equiv 1 \pmod{2^{k-1}}$, by Euler's theorem, since $\phi(2^{k-1}) = 2^{k-2}$. Suppose that in fact the order of a mod 2^{k-1} is less than 2^{k-2}, hence a proper factor of 2^{k-2} (by Chapter 6, 2.5), hence a power of 2 less than the $k - 2$ power: $a^{2^r} \equiv 1 \pmod{2^{k-1}}$ for some r with $0 \leq r \leq k - 3$. Successive squaring shows that $a^{2^{k-3}} \equiv 1 \pmod{2^{k-1}}$.

Write $a^{2^{k-3}} - 1 = 2^{k-1}b$ for an integer b.

Then $a^{2^{k-3}} + 1 = 2^{k-1}b + 2$,

and multiplying these together gives $a^{2^{k-2}} - 1 \equiv 0 \pmod{2^k}$. However, this contradicts the assumption that a is primitive mod 2^k, and this shows that the order of a is equal to $\phi(2^{k-1}) = 2^{k-2}$, as required.   □

Maybe it's time we had a positive result: we shall prove that every prime p has a primitive root. This is a mere existence theorem (in fact it actually tells us *how many* primitive roots to expect); the problem of *determining* a primitive root is an altogether different matter. There does not seem to be a really efficient way of doing this. However, see 2.11 below.

2.3 Theorem *Let p be an odd prime and $d \mid (p - 1)$. Let $\psi(d) = \#\{x: 1 \leq x \leq p \text{ and } \operatorname{ord}_p x = d\}$. Then $\psi(d) = \phi(d)$. In particular, $\psi(p - 1) = \phi(p - 1)$: the number of primitive roots mod p is precisely $\phi(p - 1)$. (Note that this holds for $p = 2$ too.)*

Proof All orders and congruences are mod p here. Note that for any x, $\operatorname{ord} x \mid (p-1)$ (Chapter 6,2.5) so we need only consider those d which divide $p - 1$. The argument is by induction on d, and the case $d = 1$ is trivial (also $d = 2$ is easy to check). Assume that for all divisors δ of $p - 1$ with $\delta < d$, we have $\psi(\delta) = \phi(\delta)$.

By 1.4 of Chapter 4, the congruence $x^d \equiv 1$ has exactly d solutions. But $x^d \equiv 1$ implies $\operatorname{ord} x = \delta$ for some $\delta \mid d$, so that $d = \sum_{\delta \mid d} \psi(\delta)$.

By the induction hypothesis, $d = \left(\sum_{\delta \mid d} \phi(\delta)\right) + \psi(d) - \phi(d)$. Using 1.7 of Chapter 6, the sum in brackets is precisely d, so that we get $\psi(d) = \phi(d)$, as required to complete the induction.   □

2.4 Exercises

(a) How many primitive roots are there mod 11? Find all of them. Do the same mod 13. For $p = 13$, verify the result of 2.3 for all divisors d of $p - 1$.

(b) Let p be a prime which is $\equiv 1 \pmod 4$. Using 2.3, why does there exist an element x of order 4 $\pmod p$? Deduce that $x^2 \equiv -1 \pmod p$. Why is it not possible for such an x to exist when p is $\equiv 3 \pmod 4$? (Solutions of equations $x^2 \equiv a \pmod p$ form the subject of Chapter 11; compare 1.5(a) in that chapter.)

(c) Let g be a primitive root mod p for a prime p of the form $8n + 1$. Why are $g^{8n} \equiv 1$ and $g^{4n} \equiv -1 \pmod p$? Let $x = g^n \pm g^{7n}$. Show that $x^2 \equiv \pm 2 \pmod p$, where $+$ and $-$ signs correspond. (Compare Chapter 11, 2.1.)

(d) Let $n = 4h$ where $h > 1$. Show that $a = 2h + 1$ gives $a^2 \equiv 1 \pmod{4h}$. Show also that $a \not\equiv \pm 1 \pmod{4h}$ and deduce that there are no primitive roots mod n from 1.3(e) above.

(e) Let g be a primitive root mod p for a prime p, so that $\pmod p$

$$1, g, g^2, \ldots, g^{p-2} \tag{1}$$

are the numbers $1, 2, \ldots, p-1$ in some order (see 1.1(b) above). Show that the differences $g - 1, g^2 - g, \ldots, g^{p-2} - g^{p-3}, 1 - g^{p-2}$ are $\pmod p$ a *cyclic* permutation of (1).

(f) Let $p = 4n + 3$ be prime and assume that $a^{(p-1)/2} = 1 \pmod p$. Show that $x = \pm a^{n+1}$ satisfies $x^2 \equiv a \pmod p$. Since this a is congruent to a square $\pmod p$ we say a is a *quadratic residue* mod p.

(g) Let $p = 8n + 5$ be prime and assume $a^{(p-1)/4} \equiv 1 \pmod p$. Verify that $x = \pm a^{n+1}$ satisfies $x^2 \equiv a \pmod p$. This a is also a quadratic residue mod p.

(h) Let p be an odd prime and suppose that m is such that, for all a with $p \nmid a$, we have $a^m \equiv 1 \pmod p$. Note that taking $a = -1$ shows that m is even. Suppose that, for some a with $p \nmid a$, we have $a^{m/2} \not\equiv 1 \pmod p$. Taking g as a primitive root mod p, show that $g^{m/2} \equiv -1 \pmod p$ and that *exactly half* of the $p - 1$ numbers $a = 1, 2, \ldots, p - 1$ satisfy $a^{m/2} \equiv 1 \pmod p$ and the other half satisfy $a^{m/2} \equiv -1 \pmod p$.

The two exercises (f) and (g) above are closely connected to the following result, which we shall make much use of in Chapter 11. Let p be an odd prime and let g be a primitive root mod p. Let $p \nmid a$ so that $a \equiv g^k \pmod p$ for some k, which is uniquely determined mod $(p-1)$ as in 1.1(a) above. Since $p - 1$ is *even* it makes sense to say that k is even or odd. Also $a^{p-1} \equiv 1 \pmod p$ by Fermat's theorem, so, p being prime (and odd!), $a^{(p-1)/2} \equiv \pm 1 \pmod p$. We then have the following:

2.5 Proposition   *With the above notation,*

(a) k is even $\Leftrightarrow a^{(p-1)/2} \equiv 1 \pmod p \Leftrightarrow a \equiv x^2 \pmod p$ for some x;

(b) k is odd $\Leftrightarrow a^{(p-1)/2} \equiv -1 \pmod p \Leftrightarrow a \not\equiv x^2 \pmod p$ for all x.

Hence we have the following, which does not mention the primitive root g :

(c) **Euler's criterion for quadratic residues:**

$x^2 \equiv a \pmod{p}$ has a solution for $x \Leftrightarrow a^{(p-1)/2} \equiv 1 \pmod{p}$.

Proof We shall prove (a); (b) is equivalent to (a) and (c) follows immediately from them. All congruences here are mod p. If $a \equiv g^{2r}$ say, then $a^{(p-1)/2} \equiv g^{(p-1)r} \equiv 1^r = 1$, and a is certainly the square of $x = g^r$. This establishes the $\Rightarrow$ implications of (a). Conversely, if $a \equiv x^2$ then writing $x \equiv g^r$ shows $a \equiv g^{2r}$ and as above $a^{(p-1)/2} \equiv 1$. To obtain the first $\Leftarrow$ of (a) assume k is odd, $a \equiv g^{2r+1}$ say; then $a^{(p-1)/2} = g^{(p-1)/2}(g^{p-1})^r \equiv g^{(p-1)/2} = h$, say. But $h^2 \equiv 1$ and h cannot be $\equiv 1$ since g is *primitive*. Hence $h \equiv -1$. $\square$

Thus, 2.4(f) gives an explicit formula for x satisfying $x^2 \equiv a \pmod{p}$, in the case when p has the form $4n+3$ and, according to Euler's criterion 2.5(c), the congruence has a solution at all. Likewise 2.4(g) gives an explicit formula when $p = 8n + 5$ and $a^{(p-1)/4} \equiv 1 \pmod{p}$. Note that Euler's criterion implies that $a^{(p-1)/4} \equiv \pm 1 \pmod{p}$; an explicit formula for x when the residue is -1 (and p has the form $8n + 5$) can also be found; see Chapter 11,2.3(e).

We shall now use the existence of primitive roots for primes to show that p^k and $2p^k$, for an odd prime p and $k \geq 1$, do have primitive roots.

2.6 Theorem *Let p be an odd prime.*

(a) *If g is a primitive root* mod p *then either g or $g + p$ is a primitive root* mod p^2.

(b) *If g is a primitive root* mod p^2 *then g is a primitive root* mod p^k *for any $k \geq 1$.*

(c) *If g is an odd (resp. even) primitive root* mod $p^k (k \geq 1)$ *then g (resp. $g+p^k$) is a primitive root* mod $2p^k$.

Remark With more careful versions of the arguments below, it can be shown that there are exactly $\phi(\phi(p^k))$ incongruent primitive roots mod p^k or mod $2p^k$. (Compare 2.3 above, which is the case $k = 1$.)

Proof

(a) Let g be a primitive root mod$p$, hence of order $p - 1$. In Chapter 6,2.6(e) we have the result: the order of g mod p^2 is either $p-1$ or $p(p-1)$. In the latter case g is also primitive mod p^2 since $\phi(p^2) = p(p - 1)$ and we are finished. In the former case we have $g^{p-1} \equiv 1 \pmod{p^2}$ (which is a fairly rare occurrence as you may have discovered in Chapter 6,2.8). Then we consider $g + p$: this is still primitive mod p (it is $\equiv g$) so as above the order of $g + p \bmod p^2$ is either $p - 1$ or $p(p - 1)$. We have only to eliminate order $p - 1$ to show that $g + p$ is primitive mod p^2.

So suppose for a contradiction that $(g+p)^{p-1} \equiv 1 \pmod{p^2}$. Expanding the left-hand side by the binomial theorem and using $g^{p-1} \equiv 1 \pmod{p^2}$ gives $\pmod{p^2}$

$$1 \equiv (g+p)^{p-1} \equiv g^{p-1} + (p-1)g^{p-2}p \equiv 1 + (p-1)g^{p-2}p.$$

Hence $(p - 1)g^{p-2} \equiv 0 \pmod{p}$ which gives $p \mid g$, an impossibility.

(b) Let g be primitive mod p^2. It is easy to check that g is primitive mod p, so let us concentrate on p^k for $k \geq 3$. We shall first prove the following by induction on $k \geq 2$: $g^{(p-1)p^{k-2}} \not\equiv 1 \pmod{p^k}$. When $k = 2$ this is so because g is primitive mod p^2, so assume it true for some $k \geq 2$. We have $g^{(p-1)p^{k-2}} = g^{\phi(p^{k-1})} \equiv 1 \pmod{p^{k-1}}$ by Euler's theorem. Write

$$g^{(p-1)p^{k-2}} = 1 + ap^{k-1} \text{where, by the induction hypothesis, } p \nmid a.$$

Thus $g^{(p-1)p^{k-1}} = (1 + ap^{k-1})^p \equiv 1 + ap^k \pmod{p^{k+1}}$, by the binomial theorem and using $k \geq 2$ to make sure subsequent terms of the expansion are divisible by p^{k+1}. (It is at this point that we need $p \neq 2$, since if $p = 2$ then the next term of the binomial expansion is divisible by p^{k+1} only for $k \geq 3$ and the induction never gets started.) Since $p \nmid a$, $1 + ap^k$ is not $\equiv 1 \pmod{p^{k+1}}$, completing the induction.

Now we prove g (a primitive root mod p^2) is primitive mod p^k, again by induction on $k \geq 2$. Assume g is primitive mod p^{k-1} for some $k > 3$. Using Chapter 6, 2.6(e) again, the order of g mod p^k is either

$$\phi(p^{k-1}) = (p-1)p^{k-2} \text{ or } p\phi(p^{k-1}) = (p-1)p^{k-1} = \phi(p^k).$$

We have only to eliminate the first of these. But it implies that $g^{(p-1)p^{k-2}} \equiv 1 \pmod{p^k}$ and that is just what we have proved false above. Hence indeed g is primitive mod p^k, completing the induction.

(c) This is essentially trivial, and depends on the fact that $\phi(2p^k) = \phi(2)\phi(p^k) = \phi(p^k)$ for any odd prime p. For suppose that g is odd and *not* primitive mod $2p^k$. Then, for some r with $0 < r < \phi(2p^k)$, we have $g^r \equiv 1 \pmod{2p^k}$. (Of course, this would not hold if g were even!) In that case certainly $g^r \equiv 1 \pmod{p^k}$ and $0 < r < \phi(p^k)$. Hence g is *not* primitive mod p^k. When the g we are given is even we simply repeat the argument with the odd number $g + p^k$ (also primitive mod p^k of course) in place of g. □

Summing up the last few results, we have:

2.7 Theorem   *The number n possesses a primitive root if and only if $n = 1, 2, 4, p^k$ or $2p^k$ for an odd prime p and $k \geq 1$.*

Note that the method of indices as in 1.2(b) above will only work for solving congruences to a modulus which has one of the very special forms in 2.7. However there do exist 'vector indices' for other values of n. A nice account of these is given in [Schoenberg (1988)].

2.8 Computing exercise   Find the smallest number g for which g is primitive mod 29, but not mod 29^2. Show that the smallest prime p for which 10 is primitive mod p, but not mod p^2, is $p = 487$. (Consulting the proof of 2.6(a), you are looking for a prime p with $10^{p-1} \equiv 1 \pmod{p^2}$.)

2.9 Exercise   It is worth seeing what the method of proof in 2.6 will achieve when $p = 2$.

(a) Show by an easy induction on k that, for a odd and $k \geq 3$, $a^{2^{k-2}} \equiv 1 \pmod{2^k}$. (Write $a^{2^{k-2}} = 1 + r2^k$ and square.)

(b) Show by induction on k that, for $k \geq 3$, $3^{2^{k-2}} \not\equiv 1 \pmod{2^k}$. (Actually this is true for $k = 2$ also, but your first step will probably be $3^{2^{k-2}} \equiv 1 \pmod{2^k}$ from (a) and this needs $k \geq 3$.)

(c) Show by induction on k that, for $k \geq 3$, the order of 3 mod 2^k is 2^{k-2}. (The result of Chapter 6, 2.6(e) is still valid for $p = 2$, so assuming the result true for k, the order mod 2^{k+1} is either 2^{k-2} or 2^{k-1}; you eliminate the former by using (b).) Of course the same argument proves that the order of 5 or 7 mod 2^k is 2^{k-2} for $k \geq 3$.

2.10 Project   Write a program to determine the smallest primitive root mod p for each prime $p < 200$ say. (Most such smallest primitive roots are quite small numbers; the one exception is $p = 191$ for which you should find 19 is the smallest primitive root. The smallest primitive root *can* be arbitrarily large, a result which is proved in Chapter 11,3.5(f).) Then use 2.6 to find primitive roots mod p^k and $2p^k$ for $k \geq 1$.

It is worth considering one method, besides brute force, which will in principle find a primitive root g for any odd prime p. The idea is to start with any initial guess a, such as $a = 2$, and, if this is not primitive, to construct other numbers with successively larger orders mod p. Eventually (and usually rather quickly) this will produce a primitive root, though not necessarily the smallest one. The idea, which is due to Gauss, uses the result of Chapter 6, 2.10, where it is shown how to find a number whose order is the least common multiple of the orders of two given numbers.

Thus, starting with say $a = 2$, calculate $a, a^2, a^3, \ldots \pmod p$ until the first one congruent to 1 is a^r, so $r = \operatorname{ord}_p a$. If $r = p - 1$ then stop! Otherwise let b be the first number with $2 < b < p - 1$ and b *not* congruent to any of the powers of a. Let s be the order of b (mod p). If $s = p - 1$ then you need look no further! In any case it is certainly true that $s \nmid r$. [Proof: if $s \mid r$ then $b^r \equiv 1$ (mod p). But $1, a, \ldots, a^{r-1}$ are pairwise incongruent solutions of $x^r \equiv 1$ (mod p) and (Chapter 4,1.4) this congruence has *exactly* r solutions, so b would have to be a power of a]. Thus $\mathrm{lcm}(r, s) > r$. Then, as in Chapter 6,2.10, we find $r = ux$, $s = vy$ with $(u, v) = 1$ and $uv = \mathrm{lcm}(r, s)$, and consider $c \equiv a^x b^y$ (mod p). This c will have order $uv = \mathrm{lcm}(r, s) > r$. If $uv = p - 1$ then we are finished, and c is a primitive root, otherwise start again but with c instead of a.

2.11 Project: construction of a primitive root mod p   Write a program to implement the above method. (If you did Chapter 6, 2.11 that will help.) For the following primes, you should obtain the primitive root indicated:

p	7	41	73	191	311	479
root	3	7	5	189	309	477

Notice that in the last three cases, -2 is primitive.

Note that the method above is impractical for large primes, if only because of storage problems: the program has to record all the powers of a until a residue of 1 is reached, and there could be as many as $p-1$ of those. You could restrict say to $p < 10^5$ and allocate this much storage space. It is only necessary to use an array of type boolean, by declaring say

```
flag:   array[0..10000] of boolean;
```

This saves space as compared with an array of integers.

When n has a primitive root, say g, then the order of g is $\phi(n)$ and, by Euler's theorem and Chapter 6,2.5, for any a with $(a, n) = 1$ we have $\mathrm{ord}_n a \mid \phi(n)$. It is natural to ask what the corresponding result is when (as usually happens), n does *not* have a primitive root. Is there a number $\lambda(n)$ to replace $\phi(n)$ here? Thus we would like $\mathrm{ord}_n a \mid \lambda(n)$ for all a coprime to n, and the existence of *some* a for which $\mathrm{ord}_n a = \lambda(n)$.

Consider for example $n = 800 = 2^5 5^2$, for which $\phi(n) = 320$. Using 2.9(a) above, $a^8 \equiv 1$ (mod 2^5) for any odd a. By Euler's theorem, $a^{20} \equiv 1$ (mod 5^2) for any a with $5 \nmid a$. Thus for any a with $(a, n) = 1$ we shall have $a^{40} \equiv 1$ (mod $2^5 5^2$), 40 being the least common multiple of 8 and 20. Of course $40 \mid 320$; we do know for sure that $a^{\phi(n)} \equiv 1$ (mod n) but the point here is that a much *lower* power of a is always 1 (mod n). Is there an element of order exactly 40? Using 2.9(c) we have that $\mathrm{ord}_{32} 3 = 8$. Since 25 has a primitive root g(e.g. $g = 2$) there exists

g with $\mathrm{ord}_{25}g = 20 = \phi(25)$. Thus we solve

$$x \equiv 3 \;(\mathrm{mod}\ 32),\ x \equiv g \;(\mathrm{mod}\ 25)$$

simultaneously by the Chinese remainder theorem (2.10 of Chapter 3). (With $g = 2$ this comes to $x = 227$.) If $x^k \equiv 1 \;(\mathrm{mod}\ 800)$ then $x^k \equiv 1$ (mod 32 and mod 25), so that, from the above orders, $8 \mid k$ and $20 \mid k$; hence $40 \mid k$ and indeed x has order exactly 40 rather than a factor of 40. With $\lambda(800) = 40$ we have that $a^{\lambda(800)} \equiv 1 \;(\mathrm{mod}\ 800)$ for all a with $(a, 800) = 1$ and there exists an element x of order exactly $\lambda(800)$. This λ is called the *minimal universal exponent* for 800.

The general case is exactly similar and we give the result in the form of exercises. Thus for each $n \geq 1$ we seek a number $\lambda(n)$ such that, for all a with $(a, n) = 1$ we have $a^{\lambda(n)} \equiv 1 \;(\mathrm{mod}\ n)$, and furthermore $\lambda(n)$ is minimal in the sense that there exists a number whose order mod n is exactly $\lambda(n)$. The existence of such a number was first proved by R. Carmichael in 1909; the definition of the λ-function is due to E. Lucas.

2.12 Exercise: minimal universal exponent $\lambda(n)$

(a) Using the results of 2.9 above show that $\lambda(2^k) = 2^{k-2}$ for $k \geq 3$. Show separately that $\lambda(4) = 2$ and $\lambda(2) = 1$. Of course $\lambda(1) = 1$ too.

(b) Let $n = 2^{n_1}p_2^{n_2}\ldots p_r^{n_r}$ be the prime-power factorization of n, where $n_1 \geq 0$ and the other n_i are > 0. Define $\lambda(n)$ to be the least common multiple of the r numbers $\lambda(2^{n_1})$ (as in (a)) and $\phi(p_i^{n_i})$ for $i = 2, \ldots, r$. (Note that, if $n > 2$, then $\lambda(n)$ is even.) Show that, for all a with $(a, n) = 1$, we have $a^{\lambda(n)} \equiv 1 \;(\mathrm{mod}\ n)$.

(c) With $\lambda(n)$ as in (b), let g_i be a primitive root mod $p_i^{n_i}$ for $i = 2, \ldots, r$. Let x be a simultaneous solution of

$$x \equiv 3 \;(\mathrm{mod}\ 2^{n_1}),\ x \equiv g_i \;(\mathrm{mod}\ p_i^{n_i})\text{ for }i = 2, \ldots, r.$$

Show that the order of x mod n is exactly $\lambda(n)$.

2.13 Project Write a program for determining $\lambda(n)$ from n. Slightly more ambitiously, determine also an element x whose order mod n is $\lambda(n)$.

2.14 Exercise: more on Carmichael numbers Using the idea of minimal universal exponent it is easy to prove the converse of 4.9 in Chapter 4.

Let $n > 2$ and let x be as in 2.12 above, that is, x has order exactly $\lambda(n)$. Suppose that $x^{n-1} \equiv 1 \;(\mathrm{mod}\ n)$. Deduce that $\lambda(n) \mid (n-1)$. Show that this implies n is odd (remember $\lambda(n)$ is even for $n > 2$). Suppose, for a contradiction, that p is a prime (> 2) and that the p^k is the highest power of p dividing n, where $k \geq 2$. Use $\phi(p^k) \mid \lambda(n)$ to show $p \mid (n-1)$. This contradiction shows that n is a product of distinct odd primes. Finally use $\phi(p) = (p-1) \mid \lambda(n)$ to show that, for each prime $p \mid n$, we have $(p-1) \mid (n-1)$.

This, together with 4.9 of Chapter 4, proves that n is a Carmichael number if and only if $n = q_1 q_2 \ldots q_k$, $k \geq 2$, where the q_i are distinct odd primes and, for each i, $(q_i - 1) \mid (n-1)$. (Note that, by 4.10(b) of Chapter 4, we must in fact have $k \geq 3$.)

9

The number of divisors d and the sum of divisors σ

Let n be an integer > 0. In this chapter we shall investigate the functions
$d(n) = \#\{x\colon 1 \leq x \leq n \text{ and } x \mid n\} =$ number of (positive) divisors of n,

$$\sigma(n) = \sum_{x \mid n} x = \text{ sum of (positive) divisors of } n.$$

Notice that $d(n)$ can be written in the slightly odd way $d(n) = \sum_{x \mid n} 1$: count one for each divisor (understood positive) of n. For example, $d(8) = 4$, the divisors being 1,2,4 and 8. Similarly $d(36) = 9$, with divisors $x = 2^i 3^j$ for i and j independently running from 0 to 2, that is $x = 1, 2, 4, 1 \cdot 3, 2 \cdot 3, 4 \cdot 3, 1 \cdot 9, 2 \cdot 9, 4 \cdot 9$. Notice that this d is *odd*; can you see why only perfect squares will give an odd value for d ?

Likewise $\sigma(8) = 1+2+4+8 = 15$, $\sigma(28) = 1+2+4+7+14+28 = 56$, $\sigma(36) = 91$. Note that squares *and* twice squares seem to give odd values for σ. Also $\sigma(28) - 28 = 28$: the sum of the divisors other than 28 itself is 28. This peculiar property, $\sigma(n) - n = n$, that is $\sigma(n) = 2n$, was considered to be particularly auspicious by the ancients, who named such numbers *perfect* numbers. Only a handful of perfect numbers are known, and they are intimately linked with Mersenne primes. No odd perfect numbers are known.

1. The function $d(n)$

The method given above for calculating $d(36)$ clearly generalizes: if

$$n = p_1^{n_1} p_2^{n_2} \dots p_k^{n_k}$$

is the prime-power decomposition of $n > 1$, then in forming a divisor of n we have $n_i + 1$ choices for the power to which p_i is included, namely $0, 1, 2, \ldots, n_i$. These choices are independent, so we have

1.1 Proposition *With the above notation,* $d(n) = (n_1 + 1)(n_2 + 1) \ldots (n_k + 1)$. *(Note than* $d(1) = 1$.)

1.2 Exercises

(a) Use this formula to prove that $d(n)$ is odd if and only if n is a perfect square.

(b) Show that, given any number $r > 1$, there exist infinitely many n with $d(n) = r$. Find all solutions n of $d(n) = 3$.

(c) Find the *smallest* numbers with the following values for d :
$d(n) = 10$, $d(n) = 25$, $d(n) = 50$, $d(n) = 100$.
[For $d(n) = 10$, for example, 10 can be written 10 or $2 \cdot 5$ or $5 \cdot 2$; choosing as small primes as possible in each case the numbers $2^9, 2^1 \cdot 3^4$ and $2^4 \cdot 3^1$ all have $d = 10$; the smallest of these is 48 so this is the answer.]

(d) Show that, if $n > 2$, then $2 \le d(n) < n$. [On no account use 1.1; just ask yourself how $d(n)$ could possibly be as large as n: for $n > 2$ can you name one x with $1 \le x \le n$ which is surely *not* a divisor of n?]

(e) Let $n_1 = d(n)$, $n_2 = d(n_1)$, $n_3 = d(n_2)$, etc. Use (d) to show that for $n > 2$ this sequence is strictly decreasing until it reaches 2 (after which it is constant).

(f) Use 1.1 to show that, if $(m, n) = 1$, then $d(mn) = d(m)d(n)$. We have met this property in connexion with Euler's ϕ-function in 1.3 of Chapter 6: it shows that d is a *multiplicative function*.

It is an easy matter to write a program for calculating d, using 1.1 and the file of primes as in 1.10 of Chapter 2. This is certainly superior to searching through all the numbers x from 1 to n and scoring 1 whenever $x \mid n$.

1.3 Program for the divisor function $d(n)$

```
PROGRAM P9_1_3;
VAR
prime, n, nsaved, total, power:   extended;
f :   text;
BEGIN
  writeln('Type the value of n');
  readln(n);
  nsaved:=n;
```

```
    (* We need to keep a note of the original value *)
    (* of n since it is continually replaced by *)
    (* n/prime later *)
  assign(f, 'PRIMES.DAT');
    (* or 'A: PRIMES.DAT' if the file is in drive A *)
  reset(f);
  total:=1;
    (* total will record the product of the nᵢ+1 *)
    (* in 1.1 *)
  WHILE (n>1) DO
    BEGIN
      read(f, prime);
      power:=1;
      WHILE (n/prime = INT(n/prime)) DO
        BEGIN
          n:=INT(n/prime);
          power:=power + 1;
        END;
      total:=total*power;
    END;
  writeln ('d(', nsaved:0:0, ')=', total:0:0);
END.
```

1.4 Computing exercises

(a) Use the above program to show that $d(242 + k)$ is the same for $k = 0$, 1, 2 and 3, and that $d(40\,311 + k)$ is the same for $k = 0$, 1, 2, 3 and 4. (In fact the latter example is, I believe, the longest such chain known!)

(b) Modify P9_1_3 so that it calculates $n_1 = d(n)$, $n_2 = d(n_1)$, etc. as in 1.2(e) above, and so that it finds the smallest i with $n_i = 2$. Make a table of values of this i for $n = 2$, 4, 8, ..., 2^{15} and for $n = 1000$, 2000, ..., $10\,000$.

(c) Find all the values of $n < 200$ for which $d(n) = d(n + 1)$. Which of these come from pairs of consecutive numbers each of which is the product of two distinct primes (making $d(n) = d(n + 1) = 4$)? Show that there are just 118 values of $n < 1000$ for which $d(n) = d(n+1)$. [It was proved by D.R. Heath-Brown in 1984 that there are infinitely many n for which $d(n) = d(n + 1)$.]

(d) There are certain values of n for which $d(m) < d(n)$ *whenever* $1 \leq m < n$. For example, $d(36) = 9$ and $d(m) \leq 8$ for $m = 1, 2, \ldots, 35$. Verify that the first few of these values of n are 1, 2, 4, 6, 12, 24, 36,

48, 60, 120, 180, 240, 360, 720, 840, 1260, 1680, 2520. Such numbers have been called *highly composite numbers* in [Ratering (1991)]. In that article he also shows that the only highly composite numbers which are divisors of all subsequent ones are 1, 2, 6, 12, 60 and 2520.

1.5 Project

This extends 1.4(c) above. Find the number of $n < 1000$ for which $d(n) = d(n + 1)$.

It is possible for n, $n + 1$ and $n + 2$ all to be products of two distinct primes, e.g. $33 = 3 \cdot 11$, $34 = 2 \cdot 17$, $35 = 5 \cdot 7$; for such n we have $d(n) = d(n + 1) = d(n + 2) = 4$. Find all $n < 1000$ for which this is the case (there are 13 such n, the largest being 921). Find all $n < 1000$ for which we just have $d(n) = d(n + 1) = d(n + 2)$.

Show that *four* consecutive integers cannot all be the product of two distinct primes (that is a (very easy) mathematical question, not a computing one!). Show that the smallest example of $d(n) = d(n + 1) = d(n + 2) = d(n + 3)$ is the one given in 1.4(a) above.

1.6 Project

It can be shown that the *average value of d(n)* for $1 \leq n \leq N$, that is $\frac{1}{N} \sum_{1}^{N} d(n)$, is approximately $\log N$ for large values of N (see for example [Hardy and Wright (1979), §18.2]; the argument is elementary and well worth following). Amend P9_1_3 to calculate this average and to calculate the difference between the average and $\log N$; make a table for suitably large N. Do your figures support the statement that the difference is approximately of the form $a + b/\sqrt{N}$ for constants a and b ?

1.7 Exercise

There is a well-known formula

$$1^3 + 2^3 + \ldots + b^3 = (1 + 2 + \ldots + b)^2.$$

Of course it follows that

$$(1^3 + 2^3 + \ldots + b_1^3)(1^3 + 2^3 + \ldots + b_2^3) \ldots (1^3 + 2^3 + \ldots + b_r^3)$$
$$= ((1 + 2 + \ldots + b_1)(1 + 2 + \ldots + b_2) \ldots (1 + 2 + \ldots + b_r))^2.$$

Use this and the formula for $d(n)$ given in 1.1 above to show the following (which is due to J. Liouville). Take any number n and let x_1, $x_2, \ldots, x_k$ be its divisors (so $k = d(n)$). Then

$$(d(x_1))^3 + (d(x_2))^3 + \ldots + (d(x_k))^3 = (d(x_1) + d(x_2) + \ldots + d(x_k))^2.$$

For example, with $n = 42$ the divisors are 1,2,3,6,7,14,21 and 42; these have d equal to 1,2,2,4,2,4,4,8 respectively, and we have

$$1^3 + 2^3 + 2^3 + 4^3 + 2^3 + 4^3 + 4^3 + 8^3 = (1 + 2 + 2 + 4 + 2 + 4 + 4 + 8)^2 = 729.$$

2. Multiplicative functions and the sum function $\sigma(n)$

Let $\mathcal{N}$ be the set of natural numbers 1, 2, 3,.... Recall that a *multiplicative function* $f\colon \mathcal{N} \to \mathcal{N}$ is one which satisfies $f(mn) = f(m)f(n)$ whenever $(m, n) = 1$. Two trivial examples of multiplicative functions are $f(n) = 1$ for all n and $f(n) = n$ for all n. Given any function $f\colon \mathcal{N} \to \mathcal{N}$, the *sum function* F of f is defined by $F(n) = \sum_{x|n} f(x)$, the sum being over all positive divisors x of n. Thus the divisor function d considered in Section 1 above is the sum function F corresponding to $f(n) = 1$ for all n while the function σ is the sum function F corresponding to $f(n) = n$ for all n. The following theorem re-proves the multiplicativity of d (see 1.2(f) above) and proves multiplicativity for σ.

2.1 Theorem *If f is multiplicative, then the sum function F of f is also multiplicative.*

Proof Let $(m, n) = 1$. We use the fact, proved in Chapter 1,2.4, that the divisors x of mn are in one-to-one correspondence with pairs of divisors x', x'' of m and n respectively. Thus

$$F(mn) = \sum f(x), \quad \text{summed over } x \mid mn$$
$$= \sum f(x'x''), \quad \text{summed over } x' \mid m \text{ and } x'' \mid n$$
$$= \sum f(x')f(x''), \quad \text{since } f \text{ is multiplicative and } (x', x'') = 1$$
$$= \left(\sum f(x')\right)\left(\sum f(x'')\right),$$
$$\text{the sums being respectively over } x' \mid m \text{ and } x'' \mid n$$
$$= F(m)F(n), \quad \text{as required.} \qquad \square$$

It is well worth carrying out the above proof in a special numerical case, e.g. $m = 10$, $n = 7$ (but keeping a general f in the proof), to convince yourself that the various expressions for $F(mn)$ really are equal!

2.2 Corollary *σ (and also d) is multiplicative.*

Proof As mentioned above, apply the theorem to $f(n) = 1$ for d and $f(n) = n$ for σ. $\qquad \square$

It is now a relatively straightforward matter to find a general formula for $\sigma(n)$ in terms of the prime decomposition of n. It is clear that, for any prime p, we have $\sigma(p) = 1 + p$. For a prime power p^k, the divisors are $1, p, p^2, \ldots, p^k$, whose sum is $\sigma(p^k) = \frac{p^{k+1}-1}{p-1}$. Hence, using 2.2, we have:

2.3 Theorem   *If the prime decomposition of* $n > 1$ *is* $n = p_1^{n_1} p_2^{n_2} \ldots$
$p_k^{n_k}$, *then*

$$\sigma(n) = \frac{p_1^{n_1+1} - 1}{p_1 - 1} \frac{p_2^{n_2+1} - 1}{p_2 - 1} \ldots \frac{p_k^{n_k+1} - 1}{p_k - 1}.$$

(Note that $\sigma(1) = 1$.)

2.4 Exercises

(a) Make a table of values of $\sigma(p^k)$ for small primes p and integers $k \geq 1$. Hence show that $\sigma(n) = 31$ has precisely the solutions $n = 2^4$ and $n = 5^2$. Find all the values of n for which $\sigma(n) = 24$, and all n for which $\sigma(n) = 28$.

(b) Let p be an odd prime, and $k \geq 1$. Show that $\sigma(p^k)$ is odd if and only if k is even. Now suppose that $\sigma(n)$ is odd. Deduce that all the powers of odd primes occurring in the prime decomposition of n are even, and hence that n has the form m^2 or $2m^2$ for some integer m.

(c) Suppose that $\sigma(n) = 3n$ and n has the form $2^k \cdot 3 \cdot 7$. Find the possible value or values of k. [A harder problem is to assume only that $n = 2^k \cdot 3 \cdot p$ for an odd prime p, and to show that p must be 7.]

(d) Let $n = 2^5 \cdot 3^3 \cdot p \cdot q$, where p and q are primes and $3 < p < q$. Suppose that $\sigma(n) = 4n$. Show that p and q are then uniquely determined and find their values.

(e) Let $k > 1$ and $a = 3 \cdot 2^k - 1$, $b = 3 \cdot 2^{k-1} - 1$, $c = 9 \cdot 2^{2k-1} - 1$. Assume that a, b and c are all prime. Writing $m = 2^k ab$ and $n = 2^k c$, show that $\sigma(m) = \sigma(n) = m + n$. Pairs m, n with $m \neq n$ satisfying this last equation are called *amicable pairs*. Note that $\sigma(m) - m = n$ and $\sigma(n) - n = m$: the sum of the divisors of m, other than m itself, is n, and vice versa.

Amicable pairs have a long history; the formula given here was known to Thabit ibn Qurra (826–901 A.D.), but he explicitly stated only the pair $220, 284$, which comes from $k = 2$ and was probably known much earlier, even to Pythagoras. The values $k = 4$ and $k = 7$ also give amicable pairs, the first noticed in the thirteenth century by Ibn al-Banna and in 1636 by Fermat, and the second by Descartes in 1638. But apart from this the Thabit rule does not seem very productive. For more information see [Wells (1986), entry 220], [Lee and Madachy (1972)], and books on the history of mathematics, such as [Kline (1972)]. A few more examples are given in 2.6(c)–(e) below.

(f) Suppose that the even number $2m$ is the sum of two distinct primes, $2m = p + q$. Show that $\sigma(pq) - pq = 2m + 1$.

It is *Goldbach's conjecture* that *every even number* > 4 *is the sum of two odd primes*, but this remains unproved although it was formulated in 1742. (Remark: Goldbach actually conjectured, in a letter to Euler, that *every number n > 5 is the sum of three primes*. Show that this is equivalent to the statement above.) The result here shows that, if Goldbach's conjecture in the slightly stronger form that *every even integer > 6 is the sum of two distinct primes* is true, then every odd number > 7 is a value of the function $s(n) = \sigma(n) - n$. Of course $6 = 3 + 3$ is not the sum of two distinct primes; all the same $s(n) = 7$ has a small solution; can you find it?

(g) Show that the equation $s(n) = 5$ has no solutions. (As usual, $s(n) = \sigma(n) - n$.) [Hint. If n is prime, then $s(n) = 1$. If p is prime, $p \mid n$, $p < n$ and $n \neq p^2$, then $s(n) \geq 1 + p + n/p$. Why does this imply that, for any given p, only values of $n \leq p(4 - p)$ need be considered?] See also 2.6(b) below.

(h) Let p be prime. Show that, if $\sigma(p^k)$ is a power of 2, then $k = 1$. [Suggestion. Assume $\sigma(p^k) = 1 + p + p^2 + \ldots + p^k = 2^s$, where $k > 1$. Deduce that $(1 + p) \mid 2^s$, so that $1 + p = 2^t$ say, and that k is odd, $k = 2r + 1$, say. Now deduce $1 + p^2 + p^4 + \ldots + p^{2r} = 2^{s-t}$, and hence $1 + p^2$ is also a power of 2. A contradiction is not far away now.] Deduce that, for any n, if $\sigma(n)$ is a power of 2, then n is a product of distinct Mersenne primes (that is primes of the form $2^m - 1$).

It is quite easy to modify program P9_1_3 above so that it calculates σ instead of d. We need, for each prime $p \mid n$, to calculate the running total of $1 + p + p^2 + \ldots$ so long as successive powers of p divide n. In the program below, each power of p (called *power_of_prime*) is calculated from the previous one by multiplication by p, and the sum (called *sum_of_powers*) is then augmented accordingly. After that prime p has been dealt with, *sigma* is updated by multiplying by the number *sum_of_powers*.

2.5 Program for the sum of divisors function σ

```
PROGRAM P9_2_5;
VAR
prime, n, nsaved, sigma:  extended;
power_of_prime, sum_of_powers:  extended;
f :  text;
BEGIN
   writeln('Type the value of n');
   readln(n);
   nsaved:=n; assign(f,'PRIMES.DAT');
   (* or 'A:PRIMES.DAT' if the file is in drive A *)
```

```
reset(f);
sigma:=1;
WHILE (n>1) DO
   BEGIN
      read(f, prime);
      power_of_prime:=1;
      sum_of_powers:=1;
      WHILE (n/prime = INT(n/prime)) DO
         BEGIN
            n:=INT(n/prime);
            power_of_prime:=power_of_prime* prime;
            sum_of_powers:=sum_of_powers + power_of_prime;
         END;
      sigma:=sigma * sum_of_powers;
   END;
   writeln ('sigma(', nsaved:0:0, ')=', sigma:0:0);
END.
```

2.6 Computing exercises

(a) **Iteration of the function** $s(n) = \sigma(n) - n$ Modify the above program so that it calculates $n_1 = s(n)$, and then takes this for the new value of n, iterating the process say m times and printing out the resulting sequence of values $n_1 = s(n)$, $n_2 = s(n_1)$, ... Check that with $n = 28$, the sequence consists entirely of 28, while with $n = 12\,496$, the values recycle after a few terms. On the other hand with $n = 100$ say, the terms are 117, 65, 19, 1. (In fact as written the program will probably produce a cycle here too, since the computer is likely to evaluate $\sigma(1) - 1 = 0$, $\sigma(0) - 0 = 1$, and so on! Can you see why the program evaluates $\sigma(0)$, which is actually undefined, as 1?) There are much longer cycle periods for some n, but the prime factors needed are likely to go off the end of the file of primes. Examples are $n = 138$ and $14\,316$.

Do all the numbers from 1 to 100 produce sequences which end in 1? What is the longest sequence produced? (It is conjectured that all such sequences go to 1 or are periodic; the smallest n for which this is in doubt is $n = 276$.)

(b) In 2.4(g) above a method is given for checking that $s(n) = 5$ has no solutions. Using the same method, how many values of n need to be checked to show that $s(n) = 52$ has no solutions? [Hint. What is the maximum value of $p(51 - p)$?] Use a program to check this and to show that 52 is the first number $k > 5$ for which $s(n) = k$ has no solutions. What is the next such $k > 52$?

(c) The result of 2.4(e) above shows that, provided three numbers a, b, c of a certain form are all prime, the numbers $m = 2^k ab$ and $n = 2^k c$ form an amicable pair. [That is, $m \neq n$ and $\sigma(m) = \sigma(n) = m + n$.] Taking $2 \leq k \leq 15$, find all values of k which give amicable pairs. [You have now found all values of $k < 20\,000$ which work, though proving this is beyond our scope here!]

(d) Despite appearances, the calculations in this question are, with a little cunning, all performable within extended precision. Let

$$a = 5^4 \cdot 7^3 \cdot 11^3 \cdot 13^2 \cdot 17^2 \cdot 19 \cdot 61^2 \cdot 97 \cdot 307$$

and let $m = a \cdot 140\,453 \cdot 85\,857\,199$, $n = a \cdot 56\,099 \cdot 214\,955\,207$. Show (using trial division) that the four large factors in m and n are in fact all prime. Show that m and n are amicable (see (c) above). [These were found in 1988 and are the smallest known where both numbers are odd and neither is a multiple of 3.]

(e) Note that, if m and n are amicable (see (c) above), then, defining $s(m) = \sigma(m) - m$, we have $s(s(m)) = m$, and similarly for n. Conversely, if $s(s(m)) = m$, then defining $n = s(m)$ we have either $m = n$ or m and n amicable. Use this and a suitable program to show that the only amicable pair with m and $n < 1000$ is $220 = 2^2 \cdot 5 \cdot 11$ and $284 = 2^2 \cdot 71$. Find the next smallest amicable pair. (Both numbers are less than 1500. This pair eluded even the great Euler, who discovered many amicable pairs in 1750. In fact it was discovered by a 16 year old Italian schoolboy, B.N.I. Paganini, in 1866.)

Needless to say there exist extensive lists of amicable pairs. [Alanen et al. (1967)] contains all pairs with the smaller number $< 10^6$; [Costello (1991)] contains amicable pairs of 'type $(i, 1)$', that is where, apart from common factors, the first number has i distinct prime factors and the second has 1. For example, $3 \cdot 5^2 \cdot 11 \cdot 31 \cdot 7 \cdot 67 \cdot 2749$ and $3 \cdot 5^2 \cdot 11 \cdot 31 \cdot 1\,495\,999$ are amicable of type $(3,1)$.

2.7 Project   Estimate the value of

$$\frac{1}{N^2} \sum_{n=1}^{N} \sigma(n)$$

for large values of N, by calculating its value for $N = 1, 2, \ldots, 500$ say. The limiting value as $N \to \infty$ is $\pi^2/12$; see for example [Hardy and Wright (1979), §18.3].

2.8 Definition: perfect numbers   A *number n is called* perfect *if*

$\sigma(n) = 2n$ *(and* deficient *(or* defective*) if $\sigma(n) < 2n$, abundant if $\sigma(n) >$
$2n$).*

For example, 6 and 28 are perfect; the next perfect number is 496. In
fact in Euclid's *Elements* (Book IX Prop 36) a formula is given which
produces even perfect numbers:

If $2^m - 1$ is prime then $2^{m-1}(2^m - 1)$ is perfect.

We leave it as an easy exercise to check this from the formula in 2.3.
The converse is due to Euler (a posthumous publication in 1849, but the
proof here is due to L.E.Dickson (1910)):

2.9 Theorem on even perfect numbers *If n is an even perfect
number then there exists m such that $2^m - 1$ is prime and $n = 2^{m-1}(2^m -$
$1)$.* (Recall that a prime of the form $2^m - 1$ is called a *Mersenne prime*.)

Proof Suppose n is an even perfect number and write $n = 2^s t$ where
t is odd. Then

$$2^{s+1}t = 2n = \sigma(n) = \sigma(2^s)\sigma(t) = (2^{s+1} - 1)\sigma(t),$$

so $2^{s+1} \mid \sigma(t)$. Write $\sigma(t) = 2^{s+1}q$ say. We proceed to show that $q = 1$.
  Suppose $q > 1$, so that $t = (2^{s+1} - 1)q$. Then q is a proper factor
of t and t has distinct factors 1, q, t (and possibly others), making
$\sigma(t) \geq 1 + q + t$. On the other hand

$$\sigma(t) = 2^{s+1}q = (2^{s+1} - 1)q + q = t + q.$$

This contradiction proves $q = 1$, giving $t = 2^{s+1} - 1$, and $\sigma(t) = t + 1$.
The last equation implies that t is prime, so $n = 2^s(2^{s+1} - 1)$ where the
second factor is prime.   □

Thus there is an intimate relationship between even perfect numbers
and Mersenne primes. The values of m making this number prime are
all prime themselves (compare 1.5(c) in Chapter 1). The first few are
$m = 2, 3, 5, 7, 13, 17, 19, \ldots$. The theorem says nothing about odd
perfect numbers; no such is known, nor is it known that none exists.
There are certainly no small ones; for example in [Brent and Cohen
(1989)] an algorithmic approach, together with quite a bit of computer
time, is used to show that there is no odd perfect number $< 10^{160}$. A
number n with $\sigma(n) = kn$ is sometimes called *k-perfect*; examples of
these have been given in 2.4(c) and (d) above.

2.10 Exercises
  (a) Let m be odd and > 1 and let $n = 2^{m-1}(2^m - 1)$. Show that
$n/2 \equiv 3$ or $4 \pmod 5$ by using such things as $2^2 \equiv -1 \pmod 5$. Deduce
that every even perfect number, in decimal notation, ends in 6 or 8.

(b) Use the standard formula for a sum of odd cubes

$$1^3 + 3^3 + 5^3 + \ldots + (2k-1)^3 = k^2(2k^2 - 1)$$

and (a) above to show that every even perfect number > 6 is a sum of consecutive odd cubes. Find the value of k for the first four even perfect numbers > 6.

(c) Use the fact that $\sum_{x|n} x = \sum_{x|n} \frac{n}{x}$ to show that, if n is perfect, then $\sum_{x|n} \frac{1}{x} = 2$, so, excluding $x = 1$, the sum of the reciprocals of the remaining divisors of n is 1. Presumably this means that a perfect number cannot have very many small divisors!

(d) Suppose that $n = p^a q^b$ where p and q are distinct odd primes. Show that $\sigma(n) < \frac{npq}{(p-1)(q-1)}$ and using $p \geq 3$, $q \geq 5$, show that this is $\leq \frac{15}{8}n$. Deduce that n cannot be perfect. (So any odd perfect number must have at least three distinct prime divisors.)

(e) Suppose that $n = p^a q^b r^c$ where p, q and r are distinct odd primes. Show that $\sigma(n) < \frac{npqr}{(p-1)(q-1)(r-1)}$. Now assume that n is perfect; deduce that $\left(1 - \frac{1}{p}\right)\left(1 - \frac{1}{q}\right)\left(1 - \frac{1}{r}\right) < \frac{1}{2}$. Show that if $p > 3$ then the left-hand side is $\geq \frac{240}{385}$ and deduce $p = 3$. Now show that $q > 5$ leads similarly to a contradiction, and from $q = 5$ deduce $r < 16$. Eliminate $r = 7, 11$ and 13 directly. Hence: any odd perfect number must have at least *four* distinct prime divisors.

2.11 Project: weird numbers   This is frankly a rather weird project, since we do not know of an efficient way to tackle it, but of course you might find one! An integer n is called *weird* if (a) $\sigma(n) > 2n$ and (b) there is *no* set of distinct divisors of n whose sum is $2n$. Check that $n = 70$ is weird. (It might be quicker to check that no sum of distinct divisors equals $\sigma(n) - 2n$ in this case.) Let n be weird and let p be a prime satisfying $p > \sigma(n)$. Prove that pn is weird (start by writing down the divisors of pn in terms of those of n). Prove that numbers of the form p^k, pq (p and q primes, $k \geq 1$) can never be weird. [In each case, show that $\sigma(n) \leq 2n$.]

Now comes the tricky part: there is just one weird number > 70 and < 900. Write a program to find it. [The problem is to find a reasonably efficient way of checking all sums of distinct divisors of n. Of course if $\sigma(n) \leq 2n$ then n can be rejected at once. Assuming that you know the divisors $n = d_1, d_2, \ldots, d_k = 1$ of n in descending order, you need to consider all sums $\sum_1^k e_i d_i$ where $e_i = 0$ or 1 for each i. Starting with all $e_i = 1$, symbolically $e = 11 \ldots 11$, consider in succession $e = 11 \ldots 10$, $e = 11 \ldots 01$, $e = 11 \ldots 00$, and so on in descending order as if the e's were binary numbers. Of course, you are looking for sums which equal exactly $2n$. It is worth missing out a few binary numbers which will certainly give sum $> 2n$ by the following short-cut: if $e_1 d_1 + e_2 d_2 + \ldots + e_r d_r > 2n$, where $e_r = 1$, then change e_r to 0, all of $e_{r+1}, \ldots, e_k$ to 1 and continue to descend down the list of binary numbers. You will find that the program quickly shows that a number is not weird, but naturally takes some time to check that the one weird number in the given range *is* weird.]

2.12* Exercises: unitary divisors and unitary perfect numbers

A *unitary divisor* x of a positive integer n is a divisor for which $(x, n/x) = 1$. Thus for example the unitary divisors of 12 are $1, 3, 4$ and 12. Let $n = p_1^{n_1} p_2^{n_2} \ldots p_k^{n_k}$ be the prime decomposition of n.

(a) Show that the unitary divisors of n are precisely divisors of the form $x = p_1^{x_1} p_2^{x_2} \ldots p_k^{x_k}$ where $x_i = 0$ or n_i for $i = 1, 2, \ldots, k$.

(b) Deduce that the *sum* $\sigma^*(n)$ of the unitary divisors of n is

$$\sigma^*(n) = (1 + p_1^{n_1})(1 + p_2^{n_2}) \ldots (1 + p_k^{n_k}).$$

(c) Show from (b) that σ^* is multiplicative: if $(m, n) = 1$ then $\sigma^*(mn) = \sigma^*(m)\sigma^*(n)$.

(d) Deduce that $\sigma^*(n)$ is odd if and only if n is a power of 2.

(e) Verify that $\sigma^*(n) = 2n$, that is n is *unitary perfect*, for the following values of n:

$6, 60, 90, 87\,360, 2^{18} \cdot 3 \cdot 5^4 \cdot 7 \cdot 11 \cdot 13 \cdot 19 \cdot 37 \cdot 79 \cdot 109 \cdot 157 \cdot 313.$

A pocket calculator is useful for the last one, but no computer is necessary! In fact [Wall (1975)] shows that these are the *first five* unitary perfect numbers.

2.13* Unitary amicable numbers   The following material is based on [Hagis (1971)]. Two integers $m > 1$ and $n > 1$ are called *unitary amicable* if $\sigma^*(m) = \sigma^*(n) = m + n$. Verify from the formula in 2.12(b) above that $114, 126$ and $1140, 1260$ are two such pairs.

(a) Using 2.12(d) show that m and n are both odd or both even.

(b) Let $m = 2^a M$, $n = 2^b N$ where $b > a > 0$, M is odd and has s distinct prime divisors, and N is odd and has t distinct prime divisors. Show that $m + n = 2^a r$ where r is odd and deduce from $\sigma^*(m) = m + n$ that $s \leq a$. Show similarly that $t \leq a$.

(c) Deduce from (b) that if $a = 1$ then M and N are powers of primes.

(d) Deduce from (b) that if M and N are perfect squares then $s = a$ and $t = a$. [Hint. For each (odd) prime p dividing M we have a factor of the form p^{2k} in the decomposition of M. Show that $2 \mid (p^{2k} + 1)$ but $4 \nmid (p^{2k} + 1)$.]

(e) Let $m = 2^a M$, $n = 2^a N$ where M and N are odd (that is $a = b$ in (b)), and assume that M and N are perfect squares. Show that $s = a+1$ and $t = a + 1$. [Hint. Writing $M = u^2$, $N = v^2$, show that u and v are odd and $2 \mid u^2 + v^2$ but $4 \nmid u^2 + v^2$.]

(f) Let $m = 2M$, $n = 2N$ where M and N are odd (so now $a = b = 1$ in (b)). Assume that $5 \nmid M$ and $5 \nmid N$. Use 2.12(c) to show that $10m$ and $10n$ are unitary amicable.

2.14 Project on unitary amicable numbers   Write a program to find unitary amicable pairs m, n (definition in 2.13 above) where m and n are even and $< 10^5$. (In fact there are no such pairs with m and n odd.)

2.15 Project: iterating the function $\sigma^*(n) - n$ In 2.6(a) we considered the iteration of the function $\sigma(n) - n$; now let $s^*(n) = \sigma^*(n) - n$ (formula for σ^* in 2.12(b) above), and let $n_1 = s^*(n), n_2 = s^*(n_1)$, etc. (Thus in the language of 2.12 and 2.13, n is unitary perfect if and only if $n_1 = n$ and m, n are unitary amicable if and only if $n_1 = m$ and $n_2 = n$.) If $n, n_1, n_2, \ldots, n_{k-1}$ are distinct but $n_k = n$ then n is said to belong to a *unitary social group* of order k. Write a program to iterate s^* and hence find the smallest n for which n belongs to a unitary social group (a) of order 3, (b) of order 5 and (c) of order 14. [The answer to (b) is $n = 1482$.]

3. Tests for primality of Mersenne numbers $2^m - 1$

In view of the close connexion between perfect numbers and Mersenne primes (see 2.9 above) it is appropriate here to state a method for determining whether a number of the form $M = 2^m - 1$ is in fact prime. First, here is a very simple fact which can help to reduce the possibilities for prime factors of M. So far as looking for prime values of M is concerned, we need only consider prime values of m (see Chapter 1,1.5(c)), and in that case the following proposition is a help. (This result was known to P. de Fermat in 1640.)

3.1 Proposition   *Let p be an odd prime, and let q be prime. Suppose that $q \mid (2^p - 1)$. Then q has the form $1 + 2kp$ for an integer k. Further, all divisors of $2^p - 1$, whether prime or not, have this form.*

Proof   By Fermat's theorem, $q \mid (2^{q-1} - 1)$, so that $q \mid (2^{q-1} - 1, 2^p - 1) = 2^{(q-1,p)} - 1$ by Chapter 1,3.13(c). Hence $(q - 1, p) = p$, rather than the only other possibility which is 1. This gives $q - 1 = rp$ but p is odd and $q - 1$ is even so r must be even, $r = 2k$ say. The last statement follows, since if $q_1 \equiv 1$ and $q_2 \equiv 1 \pmod{2p}$, then $q_1 q_2 \equiv 1 \pmod{2p}$.

Taking $p = 31$ this tells us that any prime divisor of $2^{31} - 1$ must be of the form $1 + 62k$, so the primality of $2^{31} - 1$ can be checked quickly by trial division by numbers of this special form.

3.2 Computing exercise Write a program to find divisors of numbers $2^p - 1$ by trial division by numbers of the form $2kp + 1$. Of course, if no divisor is found which is $\leq \sqrt{2^p - 1}$ then $2^p - 1$ is in fact prime. Note that all divisors are found; for example with $p = 29$ the divisors come out as $233, 1103, 2089, 256\,999, \ldots$ and this last one is 233×1103. It will take the program a long time to test all numbers of the form $2kp + 1 < 2^p - 1$, but in fact $2^{29} - 1 = 233 \cdot 1103 \cdot 2089$ so the prime factorization is found reasonably quickly. (Why must these initial factors be prime?) What is the largest value of p for which the program reliably finds prime factors? What happens if you input nonprime values of p ? Does the program find any/all/some factors of $2^p - 1$? Can you explain? (Try $p = 9$ and $p = 15$ for a start.)

It will be clear that the method of 3.2 will be very slow to prove $2^p - 1$ is prime for a large prime p. There is a remarkable primality test, specifically designed for Mersenne numbers, proved by D.H. Lehmer in 1932 (with a much simplified proof in 1934), and based on ideas of E. Lucas (1878). This test is very efficient, and is in fact routinely used, together with multiprecision arithmetic routines, for finding large Mersenne primes. (Indeed calculations of this kind are sometimes used to test the correct working of newly installed computers, as for example with the 65 050 digit prime $2^{216\,091} - 1$ discovered with a Cray XMP supercomputer in September 1985.) The Mersenne primes are fairly thinly seeded; for example there is none between $2^{127} - 1$, which has 39 digits, and $2^{521} - 1$, which has 157 digits. In 1989 the unexpected discovery was made (by W. Colquitt and L. Welsh in Texas) that there was a 'lost Mersenne prime' in between two previously known ones, namely $2^{110\,503} - 1$, between $2^{86\,243} - 1$ and $2^{132\,049} - 1$, both of which were discovered in 1983. In 1992 a Cray-2 supercomputer at the Atomic Energy Authority's Harwell Laboratory discovered the new Mersenne prime $2^{756\,839} - 1$. See *Nature*, 26 March 1992, p. 283. (In *Science*, 7 January 1994, p. 27, a new prime is announced: $2^{857\,433} - 1$.)

The test is as follows; we shall prove it one way round below and the other way round in Chapter 11, 3.6.

3.3 Lucas–Lehmer test for Mersenne primes *Let* $M_p = 2^p - 1$, *where* p *is an odd prime. Let* $r_1 = 4$ *and, for* $k \geq 2$, *let* $r_k \equiv r_{k-1}^2 - 2$ (mod M_p). *Then* M_p *is prime if and only if* $r_{p-1} \equiv 0$ (mod M_p).

Since calculations need to be done mod M_p, we cannot achieve very much with this test using only extended precision, but it is worth using the test to see how fast it is within this range. Note that it gives no clue at all to the factors of M_p when M_p is *not* prime.

3.4 Computing exercise   Write a program to implement the Lucas–Lehmer test for values of p which make M_p within extended precision. The largest prime value of M_p which can be obtained this way is M_{31}.

We conclude with a proof of the more 'interesting' part of the Lucas–Lehmer test, namely that *if* the condition $r_{p-1} \equiv 0 \pmod{M_p}$ holds, *then* M_p is prime. (For the other half, see Chapter 11, 3.6.) The proof given here was found by J.W. Bruce [Bruce (1993)], and is one of a long line of ever simpler proofs of the theorem (an ancestor in this respect is [M.I. Rosen(1988)]). The proof nevertheless does use a minute amount of group theory, explained below, so it is certainly to be regarded as optional reading. It also uses irrational numbers in a rather unexpected way; see [Ribenboim (1988), p. 76] for an explanation of how these numbers fit into the more general context of Lucas sequences.

*Proof of 'if' part of 3.3   Let $\omega = 2+\sqrt{3}$, $\overline{\omega} = 2-\sqrt{3}$. Thus $\omega\overline{\omega} = 1$ and it is easy to check by induction that, in the notation of 3.3:   □

3.5 Lemma $r_k = \omega^{2^{k-1}} + \overline{\omega}^{2^{k-1}}$

It follows from this that, if $M_p \mid r_{p-1}$, then, for some integer R, we have

$$\omega^{2^{p-2}} + \overline{\omega}^{2^{p-2}} = RM_p.$$

Multiplying this identity by $\omega^{2^{p-2}}$ and using $\omega\overline{\omega} = 1$ gives

3.6 $\omega^{2^{p-1}} = RM_p\omega^{2^{p-2}} - 1$

and squaring gives

3.7 $\omega^{2^p} = (RM_p\omega^{2^{p-2}} - 1)^2.$

Assume for a contradiction that M_p is composite, and choose a prime divisor q with $q^2 \le M_p$. Clearly $q > 2$.

We now introduce the very elementary ideas from group theory used in the rest of the proof.

3.8 Lemma   *Let X be a set with a binary operation which is associative and has an identity. Then the set X^* of invertible elements in X forms a group.*

Proof of 3.8   Clearly the identity $1 \in X^*$, so we have a nonempty set. We have to show that X^* is closed under the binary operation (it is clear that X^* is closed under the taking of inverses). But if x_1, x_2

are elements of X^* with inverses x_1^{-1}, x_2^{-1}, then $x_1 x_2$ is invertible with inverse $x_2^{-1} x_1^{-1}$, so $x_1 x_2 \in X^*$. $\square$

3.9 Lemma   *If G is a finite group then the order of an element is at most the order of the group. If $x \in G$ and $x^r = 1$, then the order of x divides r.* [Of course the *order* of G is the number of elements of G and the *order* of x means the smallest $k > 0$ for which $x^k = 1$.]

Proof of 3.9   Let $x \in G$ and consider $1 = x^0, x^1, x^2, x^3, \ldots, x^n$ where n is the order of G. There are $n+1$ elements of G here, so two (at least) must coincide, say $x^u = x^v$, with $0 \le u < v \le n$. Then $x^{v-u} = 1$, and $1 \le v - u \le n$, so that x has order $\le v - u \le n$. This proves the first assertion. (We do not need the stronger result, usually called *Lagrange's theorem*, that the order of x *divides* the order of G.) $\square$

For the second assertion (which is closely related to 2.5 of Chapter 6), write s for the order of x and assume $x^r = 1$. Let $r = as + b$ where $0 \le b < r$. Now $1 = x^r = x^{as+b} = (x^s)^a x^b = x^b$. If $b > 0$ we have a smaller power of x than x^r yielding the identity, contradicting the definition of order of an element. Hence $b = 0$ and $s \mid r$, as required.

For the proof of 3.3 we now let $\mathcal{Z}_q$ denote $\{0, 1, \ldots, q - 1\}$ under the operation of multiplication modulo q, and let X denote $\{a + b\sqrt{3} : a, b \in \mathcal{Z}_q\}$, with the induced operation given by $(a + b\sqrt{3})(c + d\sqrt{3}) = ac + 3bd + (ad + bc)\sqrt{3}$, reducing the coefficients modulo q. This operation is associative (and commutative) with identity $1 = 1 + 0\sqrt{3}$. Let X^* denote the group (see 3.8) of invertible elements of X. Since X has q^2 elements and 0 is certainly *not* invertible, X^* has at most $q^2 - 1$ elements and by 3.9 the order of any element of X^* is $\le q^2 - 1$.

Now consider $\omega = 2 + \sqrt{3}$ as an element of X. Note that $\omega \in X^*$ since $\omega \bar{\omega} = 1$. Since $q \mid M_p$ it follows that, in the notation of 3.6 and 3.7, $R M_p \omega^{2^{p-2}}$, when expanded out as an element of X, is 0. So 3.6 and 3.7 say that, in X,

$$\omega^{2^{p-1}} = -1 \text{ and } \omega^{2^p} = 1.$$

Using 3.9, the order of ω is a factor of 2^p, hence a power of 2, ord $\omega = 2^k$ say with $k \le p$. But if $k < p$ then successive squaring shows that $\omega^{2^{p-1}} = 1$, which is false. Hence the order is 2^p. (This argument may be familiar if you tried Chapter 6,2.6(h).) Using 3.9 again we have $2^p \le q^2 - 1$. However $q^2 - 1 \le M_p - 1 = 2^p - 2$ so we have a contradiction. This completes the proof of the 'if' part of the Lucas–Lehmer test 3.3. $\square$

10

Continued fractions and factoring

We have devoted some attention in previous chapters to the problem of proving that numbers are prime, or deducing that they are composite without necessarily having any idea what their factors are. In this chapter we shall describe one method for solving the harder problem of factoring a large number. As in previous chapters, in order to keep the programs self-contained and independent of multiprecision arithmetic, we shall stay within extended precision. Although this means that we cannot attempt state-of-the-art numbers with 100 digits, the idea of the method is conveyed very well by more modest examples.

In fact the method uses some very beautiful mathematics concerning continued fractions, which have been studied for centuries because of their ability to approximate irrational numbers with, in a sense, optimal accuracy. The 'continued fraction expansion' of $\sqrt{2}$ goes back to R. Bombelli (1572) but the first systematic treatment of continued fractions was done by J. Wallis (1695), who also gave them their name. Euler used them extensively from 1737. One of the most thorough treatments in English is in Volume 2 of a classic textbook called *Algebra*, by G. Chrystal (first published 1886, the seventh edition reprinted as [Chrystal (1964)]), and the beginnings of the theory are treated in [Hardy and Wright (1979), Chapter X].

We shall stick rather closely here to the properties of continued fractions which are most germane to the method of factoring. We have collected some of the more delicate proofs into Section 3. In Section 4 we apply the method of continued fractions to the solution of a non-linear equation called Pell's equation, and in Section 5 we go on to the factoring method.

The idea behind the factoring method is extremely simple, and has been mentioned before, in Chapter 3, 1.5. Namely, if $x^2 \equiv y^2 \pmod{n}$ but $x \not\equiv \pm y \pmod{n}$, then the greatest common divisor $(x + y, n)$ is a proper factor of n. What the method does is to provide a more or less systematic method of finding x and y with $x^2 \equiv y^2 \pmod{n}$; with luck, some of them will also satisfy $x \not\equiv \pm y \pmod{n}$ and yield a proper factor. Numbers which are congruent to squares, modulo n, are called *quadratic residues* modulo n, and the method can be refined by using properties of quadratic residues. We shall meet these properties in the next chapter.

The factoring method we shall describe first appeared in [Morrison and Brillhart (1975)]; see also [Riesel (1985), Chapter 5].

1. Continued fractions: initial ideas and definitions

First, a very simple example. The Euclidean algorithm applied to the numbers 67 and 28 is

$$67 = 28 \cdot 2 + 11$$
$$28 = 11 \cdot 2 + 6$$
$$11 = 6 \cdot 1 + 5$$
$$6 = 5 \cdot 1 + 1$$
$$5 = 1 \cdot 5 + 0.$$

Dividing the first line through by 28 gives $\frac{67}{28} = 2 + \frac{11}{28}$. The second line similarly gives an expression for $\frac{28}{11}$; inverting this gives

$$\frac{67}{28} = 2 + \cfrac{1}{2 + \frac{6}{11}}.$$

Similarly $\frac{11}{6} = 1 + \frac{5}{6}$ from the third line so we can substitute for $\frac{6}{11}$. One more substitution of this kind from the fourth line of the algorithm produces the final answer, called a *continued fraction*:

$$\frac{67}{28} = 2 + \cfrac{1}{2 + \cfrac{1}{1 + \cfrac{1}{1 + \frac{1}{5}}}}$$

That is almost the last time we shall use such cumbersome notation; obviously some abbreviation is needed and we shall follow convention in writing the above as

$$\frac{67}{28} = [2, 2, 1, 1, 5].$$

Note that since $5 = 4 + \frac{1}{1}$ the continued fraction could be written $[2, 2, 1, 1, 4, 1]$ although this would come less naturally from the Euclidean algorithm. This tiny ambiguity will not cause problems.

The very last time we shall use the cumbersome notation is to say that

$$a_0 + \cfrac{1}{a_1 + \cfrac{1}{a_2 + \cfrac{1}{a_3 + \cdots + \frac{1}{a_n}}}}$$

is written $[a_0, a_1, a_2, a_3, \ldots, a_n]$.

Clearly from the Euclidean algorithm every positive rational number can be written in this way with the *a*s all positive integers: if the *quotients* in the Euclidean algorithm for (a, b) are $q_1, q_2, \ldots, q_n$, as in Chapter 1,3.3, then $\frac{a}{b} = [q_1, q_2, \ldots, q_n]$. Conversely, if the *a*s are positive integers, then the above expression is certainly a rational number.

We have the opportunity here to do an interesting kind of approximation to a/b: just forget the last few entries in the continued fraction. If we do this to $67/28$, we get, by forgetting successively fewer entries,

$$[2] = 2, \ [2, 2] = 2.5, \ [2, 2, 1] = 2.\overline{3}, \ [2, 2, 1, 1] = 2.4,$$

$$[2, 2, 1, 1, 5] = 2.392\,857\ldots,$$

the last being the exact value $67/28$. These 'successive approximations' are called the *convergents* of the continued fraction; more generally $[a_0, a_1, \ldots, a_k]$, for $0 \leq k \leq n$, are the *convergents* of $[a_0, a_1, \ldots, a_n]$. We shall shortly give an efficient way of working out convergents.

Going back to our initial example of $67/28$, recall that the division used in the Euclidean algorithm guarantees that the remainders 11, 6, 5, 1, 0 are strictly less than the respective quotients 28, 11, 6, 5, 1. Thus in $\frac{67}{28} = 2 + \frac{11}{28}$, for example, the fraction $\frac{11}{28}$ is less than 1, and 2 is the integer part of $\frac{67}{28}$. We write, for any positive rational starting number x_0 which is not an integer,

$$x_0 = a_0 + \frac{1}{x_1}$$

where $a_0 = [x_0]$ and the fractional part, which is < 1, is written in the form $1/x_1$, so that $x_1 > 1$. (If x_0 is an integer then we just stop at $x_0 = a_0$.) Assuming $x_1 > 1$ and is not an integer, we can continue with

$$x_1 = a_1 + \frac{1}{x_2}$$

where $a_1 = [x_1]$ and $x_2 > 1$. In the initial example this step appeared as $\frac{28}{11} = 2 + \frac{1}{11/6}$, so that $x_2 = 11/6$. The numbers a_i produced this way are just the quotients in the Euclidean algorithm, which stops when a

number x_n is produced which is an integer: a_n is this integer and no
further steps occur.

1.1 General procedure for generating a continued fraction

Generally, suppose that we start with a positive number x_0 and produce
successive integers a_i by the above process. That is, for $i = 0, 1, 2, \ldots$
we

(∗)   take integer part: $a_i = [x_i]$, and,

   if x_i is not itself an integer subtract and invert:

   $x_i - a_i = 1/x_{i+1}$ so that $x_{i+1} = 1/(x_i - a_i)$,

   and return to (∗)

   otherwise terminate the process.

   Then the x are called the *complete quotients* and the integers a_i, which
satisfy $a_i \geq 1$ for $i \geq 1$ and $a_0 \geq 0$, are called the *partial quotients*
associated to x_0. As above, $[a_0, a_1, a_2, \ldots]$ is called the *continued fraction*
for x_0.

   We shall be interested shortly in curtailing the continued fraction at
some point, giving $[a_0, a_1, \ldots, a_n]$ say, which is *defined* by the long and
cumbersome expression written out above. To *evaluate* this it appears
that we shall have to start at the bottom (at a_n) and work our way up.
However we shall see in 1.2 that there is a simple iterative procedure
for calculating this expression from the top (a_0) down. For the present
we shall not worry about the 'convergence' of the continued fraction; in
Section 3 we take up the question as to when

$$\lim_{n \to \infty} [a_0, a_1, a_2, \ldots, a_n]$$

necessarily exists.

   There is one very useful trick which is worth knowing. In the expression $x_0 = [a_0, a_1, a_2, \ldots, a_k]$ it is not actually necessary that the a_i are
integers for the expression to make sense. Thus we can write

$$x_0 = [a_0, x_1], \quad \text{meaning } x_0 = a_0 + \frac{1}{x_1},$$

$$x_0 = [a_0, a_1, x_2], \quad \text{meaning } x_0 = a_0 + \frac{1}{a_1 + \frac{1}{x_2}},$$

(remember $x_1 = a_1 + 1/x_2$), and so on. For example we have

$$\frac{67}{28} = [2, \frac{28}{11}] = [2, 2, \frac{11}{6}] = [2, 2, 1, \frac{6}{5}] = [2, 2, 1, 1, 5].$$

More generally we have, for any continued fraction $x_0 = [a_0, a_1, a_2, \ldots]$

with complete quotients $x_0, x_1, x_2, \ldots$ calculated as in 1.1 above,

$$x_0 = [a_0, a_1, \ldots, a_{k-1}, x_k], \tag{1}$$

for $k = 1, 2, 3, \ldots$. Here, x_k can be thought of as the result of working out $[a_k, a_{k+1}, \ldots]$ as a continued fraction in its own right.

A simple case of (1) is the observation that we can group the last two terms in a continued fraction:

$$[a_0, a_1, a_2, \ldots, a_{k-1}, a_k, a_{k+1}] = [a_0, a_1, a_2, \ldots, a_{k-1}, a_k + \tfrac{1}{a_{k+1}}]. \tag{2}$$

This observation provides the key ingredient in the proof of the following simple recurrence formula for the convergents of a continued fraction. Let $[a_0, a_1, a_2, \ldots]$ be a continued fraction, and let c_k denote the kth convergent $[a_0, a_1, \ldots, a_k]$. We write $c_k = p_k/q_k$, where the numerator p_k and the denominator q_k are worked out formally in terms of $a_0, a_1, \ldots, a_k$, just manipulating fractions. Thus for example

$$c_0 = \frac{a_0}{1} = \frac{p_0}{q_0} \text{ , so } p_0 = a_0, q_0 = 1;$$

$$c_1 = a_0 + \frac{1}{a_1} = \frac{a_0 a_1 + 1}{a_1}, \text{so } p_1 = a_0 a_1 + 1, q_1 = a_1;$$

$$c_2 = a_0 + \frac{a_2}{a_1 a_2 + 1} = \frac{(a_0 a_1 + 1)a_2 + a_0}{a_1 a_2 + 1}, \text{ so}$$

$$p_2 = (a_0 a_1 + 1)a_2 + a_0, q_2 = a_1 a_2 + 1.$$

1.2 Theorem   *With the above notation, for $k \geq 1$,*

$$p_{k+1} = a_{k+1} p_k + p_{k-1}; \ q_{k+1} = a_{k+1} q_k + q_{k-1}.$$

In fact, if we define $p_{-2} = 0$, $q_{-2} = 1$, $p_{-1} = 1$, $q_{-1} = 0$, then the recurrence relation holds for all $k \geq -1$

Note that this is a purely formal algebraic result: it holds whether the a are nonzero integers, rational numbers or indeed real or complex numbers.

Proof   The proof is by induction on $k \geq 1$, and the case $k = 1$ is covered by the formula above for c_2. So assume that the recurrence holds for p_r and q_r with $1 \leq r \leq k$, and all continued fractions with at least $k+1$ entries. We must now verify the recurrence for p_{k+1} and q_{k+1}. Using the observation (2) above, we write the $(k+1)$th convergent

$$p_{k+1}/q_{k+1} = [a_0, a_1, a_2, \ldots, a_{k-1}, a_k, a_{k+1}]$$

$$= [a_0, a_1, a_2, \ldots, a_{k-1}, a_k + \frac{1}{a_{k+1}}].$$

The crucial thing is that the two continued fractions here have *the same* entries up to and including the kth entry a_{k-1}, so they have *the same* convergents p_i/q_i for $i = 0, 1, \ldots, k - 1$. Furthermore, the second continued fraction has one fewer entry than the first, so comes under the induction hypothesis. Using the induction hypothesis gives

$$c_{k+1} = \frac{p_{k+1}}{q_{k+1}} = \frac{\left(a_k + \frac{1}{a_{k+1}} \right) p_{k-1} + p_{k-2}}{\left(a_k + \frac{1}{a_{k+1}} \right) q_{k-1} + q_{k-2}}$$

$$= \frac{a_{k+1}(a_k p_{k-1} + p_{k-2}) + p_{k-1}}{a_{k+1}(a_k q_{k-1} + q_{k-2}) + q_{k-1}}.$$

Using the induction hypothesis again, this reduces to

$$c_{k+1} = \frac{a_{k+1} p_k + p_{k-1}}{a_{k+1} q_k + q_{k-1}},$$

which is exactly the form we need to complete the induction. The last statement of the theorem is easy to verify directly.   □

1.3 Exercises

(a) Verify the following continued fractions, complete quotients x (to two decimal places) and convergents c:

$$24/13 = [1, 1, 5, 2]; x : 1.18, 5.50, 2.00; c : 1, 2, 11/6, 24/13;$$

$$39/12 = [3, 4]; x : 4.00; c : 3, 39/12;$$

$$135/28 = [4, 1, 4, 1, 1, 2]; x : 1.22, 4.60, 1.67, 1.50, 2.00;$$

$$c : 4, 5, 24/5, 29/6, 53/11, 135/28.$$

Verify that the values of the convergents are alternately above and below the true starting value, and that they get successively nearer to this value.

(b) Show by induction on k that, with the ps and qs defined recursively as in 1.2, starting with $p_0 = a_0$, $q_0 = 1$, $p_1 = a_0 a_1 + 1$, $q_1 = a_1$ we have

$$p_k q_{k-1} - p_{k-1} q_k = (-1)^{k-1}, \text{ for } k \geq 1.$$

Verify that this holds in the numerical cases of (a).

Note that it shows that, in the case when all the a in the continued fraction are nonzero integers, the numerator and denominator of the kth convergent are always *coprime*. (For any common factor would divide $(-1)^{k-1}$.) The equation also shows $(p_0, q_0)(= (a_0, 1)) = 1$. Thus in this case *the convergents are always fractions in their lowest terms*.

1.4 Computing exercises

(a) Write a program to start with a number x_0 (perhaps input as a fraction a/b) and work out the successive partial quotients a_i and complete quotients x_i as above, by the simple process of taking integer parts, subtracting and inverting, so long as the complete quotient is not an integer. Try it on the following numbers: $x_0 = 10/3$, $x_0 = 12/5$, $x_0 = 67/28$. You may find, with luck, that the first two do terminate, producing the correct partial quotients (3,3 for 10/3 and 2,2,2 for 12/5) followed by one or more preposterously large values. These are caused, of course, by the fact that $x_i - a_i$ is not exactly 0, so rather than the program terminating the next x is extremely large. You will probably find that 67/28 does not appear to terminate at all: after the correct partial quotients 2,2,1,1,5 there is a pretty random assortment of numbers, some huge and others of ordinary size. It is clear that this is no way to calculate continued fractions!

(b) Now use a Euclidean algorithm program such as P1_3_5 to do the job properly: starting with a rational number $x_0 = a/b$ find the entries a_i in the continued fraction for x_0. The entries are the *quotients* in the algorithm so you will need to print them out. Print out also the complete quotients x_i. Verify that the continued fraction and partial quotients for 67/28 are as stated in the text above and that $152\,344\,421/75\,336\,901$ has continued fraction $[2, 45, 10, 1, 1, 62, 55, 3, 3, 2]$, with complete quotients (to two decimal places) 45.10, 10.50, 1.98, 1.02, 62.02, 55.30, 3.29, 3.50 and 2.00.

(c) Write a program to work out the convergents of $x_0 = a/b$, using the recurrence in Theorem 1.2 and the Euclidean algorithm as in (b) above. Of course you need to remember the last *two* numerators p and denominators q, and to initialize the first two values of p and of q as in the statement of 1.2. For instance you could use p and q to denote the current values, pp and qq to denote the previous ones and ppp and qqq to denote the ones before that. Then you initialise as

$$ppp:= 0; \quad pp:= 1; \quad qqq:= 1; \quad qq:= 0;$$

and, once the new (current) values p, q have been calculated you use

$$ppp:= pp; \quad pp:= p; \quad qqq:= qq; \quad qq:= q;$$

to get ready for the next iteration.

What are the convergents of the fraction in (b) above? Check that they are alternately above and below the true value, and converge on it.

2. The continued fraction for $\sqrt{n}$

Our main interest in continued fractions is in the special case where we start not with a rational number but with $\sqrt{n}$, where n is an integer which is not a perfect square. In this case we find that the continued fraction fails to terminate, much as the decimal expansion of an irrational number fails to terminate. However, there is one enormous difference: the decimal expansion does not recur either, whereas, remarkably, the continued fraction expansion *does* recur. It is numbers associated with the continued fraction of $\sqrt{n}$ that will give us solutions to the congruence $x^2 \equiv y^2 \pmod{n}$, and these, as noted above, lead to a proper factor of n if luck is with us.

Consider for example $x_0 = \sqrt{6}$. We try to follow the method of Section 1 above to find successive partial quotients a_i, which will be positive integers, and complete quotients x_i, which will be real numbers. It is essential to keep complete accuracy (see 1.4(a) above) so we work in terms of $\sqrt{6}$ in the calculation.

$$x_0 = a_0 + \frac{1}{x_1} : a_0 = [\sqrt{6}] = 2 \text{ so } x_1 = \frac{1}{\sqrt{6} - 2} = \frac{2 + \sqrt{6}}{2}.$$

It is worth recalling 4.3(h) of Chapter 1: for integers P and Q, with $Q > 0$, we have $\left[\frac{P + x}{Q}\right] = \left[\frac{P + [x]}{Q}\right]$. So when working out $[x_1]$ we can replace the $\sqrt{6}$ by $[\sqrt{6}] = 2$.

$$x_1 = a_1 + \frac{1}{x_2} : a_1 = [x_1] = 2 \text{ so } x_2 = \frac{2}{-2 + \sqrt{6}} = \frac{4 + 2\sqrt{6}}{2} = 2 + \sqrt{6}.$$

Note that a factor of 2 is cancelled in the last step. This is not as trivial as it sounds: without such cancellation the coefficients become enormous after a few iterations.

$$x_2 = a_2 + \frac{1}{x_3} : a_2 = [x_2] = 4 \text{ so } x_3 = \frac{1}{\sqrt{6} - 2}.$$

From this point on, the calculation will recur, for $x_3 = x_1$. The calculation of the continued fraction for $\sqrt{6}$ therefore produces, at least on the level of formal algebraic manipulations,

$$\sqrt{6} = [2, 2, 4, 2, 4, 2, 4, \ldots],$$

which can be written $[2, \overline{2, 4}]$ to show the recurrence. There is a slight worry as to whether this *infinite continued fraction* converges to $\sqrt{6}$ in the sense that the convergents p_k/q_k converge to $\sqrt{6}$. But let us leave that question for the moment (see Section 3 below) and try to do the same calculation as above for a general $\sqrt{n}$ where n is not a perfect square.

2.1 Exercise Carry out the calculation as above in the cases $n = 8$ and $n = 7$, verifying that $\sqrt{8} = [2, \overline{1, 4}]$ and $\sqrt{7} = [2, \overline{1, 1, 1, 4}]$.

All the x_i in the calculation of the continued fraction for $\sqrt{n}$ for $n = 6$ (and also for $n = 7$ and $n = 8$ if you did those) turn out to have the form (possibly after cancellation) $\frac{P + \sqrt{n}}{Q}$ for integers P and Q: note the coefficient 1 in front of $\sqrt{n}$. In fact this holds in general; the following result shows this and also provides a very convenient iterative method for finding the successive Ps and Qs and for finding the terms a_i in the continued fraction expansion $\sqrt{n} = [a_0, a_1, a_2, a_3, \ldots]$.

2.2 Theorem   *The complete quotients in the continued fraction expansion of $\sqrt{n}$ (where n is a positive integer which is not a perfect square) are*

$$x_k = \frac{P_k + \sqrt{n}}{Q_k},$$

where P_k and Q_k are integers with $Q_k \neq 0$ determined by

$$P_0 = 0, \; Q_0 = 1, \; P_{k+1} = a_k Q_k - P_k, \; Q_{k+1} = \frac{n - P_{k+1}^2}{Q_k},$$

$$k = 0, 1, 2, \ldots \; where \; a_k = [x_k].$$

Some readers may prefer to skip the induction argument and proceed straight to the program and examples which follow.

Proof   Let n be given. Assume that, for some $k \geq 0$, we have $x_k = \frac{P_k + \sqrt{n}}{Q_k}$ for integers P_k and Q_k, with $Q_k \neq 0$, and that furthermore $Q_k \mid (n - P_k^2)$. Note that all this certainly holds for $k = 0$. Let $a_k = [x_k]$. As usual, x_{k+1} is defined by $x_{k+1} = \frac{1}{x_k - a_k}$; note that x_k cannot be an integer since $\sqrt{n}$ is irrational, so the denominator here cannot be 0. Substituting for x_k and rationalizing the denominator give

$$x_{k+1} = \frac{Q_k(a_k Q_k - P_k + \sqrt{n})}{n - (a_k Q_k - P_k)^2}.$$

Now $Q_k \mid (n - (a_k Q_k - P_k)^2)$, by the assumption about Q_k; hence this can be rewritten

$$x_{k+1} = \frac{P_{k+1} + \sqrt{n}}{Q_{k+1}},$$

provided we define P_{k+1} and Q_{k+1} as in the statement of the theorem. Note that Q_{k+1} is an integer, by the property of Q_k just mentioned, and it cannot be zero since n is not the square of an integer. To complete the induction we need only check that $Q_{k+1} \mid (n - P_{k+1}^2)$. But from the formula for Q_{k+1} we have $Q_{k+1} Q_k = n - P_{k+1}^2$, so this follows at once.   □

Note the cancellation of Q_k from numerator and denominator in the expression for x_{k+1}. This is what produces the coefficient 1 in front of $\sqrt{n}$; without this cancellation the coefficients would rapidly exceed the wordsize of the computer.

2.3 Exercise   Use $Q_{k-1} = (n - P_k^2)/Q_k$, $Q_{k+1} = (n - P_{k+1}^2)/Q_k$ to show that $Q_{k+1} - Q_{k-1} = a_k(P_k - P_{k+1})$, for $k \geq 1$. This can be used to find each Q from the one *two* steps before, and without having to do a division.

Numerical examples are very easy to carry through, using the formulas of 2.2. For example, let $n = 14$. It is convenient to set out the calculation in the form of a table which reflects the order in which the numbers are calculated, namely P_0, Q_0, x_0, a_0, P_1, Q_1, x_1, a_1,.... Remember that when calculating $a_k = [x_k]$ we can replace $\sqrt{n}$ by $[\sqrt{n}]$ this is 4.3(h) of Chapter 1, which was used above in calculating the case $n = 6$.)

$n = 14$	k	P	Q	x		a
	0	0	1	$\sqrt{14}$		3
	1	3	5	$\frac{3+\sqrt{14}}{5}$		1
	2	2	2	$\frac{2+\sqrt{14}}{2}$		2
	3	2	5	$\frac{2+\sqrt{14}}{5}$		1
	4	3	1	$3 + \sqrt{14}$		6
	5	3	5	same as	$k = 1$:	recurrence starts.

2.4 Exercises

(a) Work through the examples of 2.1 again using this method.

(b) Let $n = d^2 + 1$, where $d \geq 1$. Show that $\sqrt{n}$ has continued fraction expansion $[d, \overline{2d}]$.

(c) Let $n = d^2 - 1$, where $d \geq 1$. Show that $\sqrt{n}$ has continued fraction expansion $[d - 1, \overline{1, 2d - 2}]$.

(d) Let $n = d^2 - d$, where $d \geq 2$. Show that $\sqrt{n}$ has continued fraction expansion $[d, \overline{d, 2d}]$.

Notice that just before recurrence starts we have $Q_k = 1$. In fact, suppose that $Q_k = 1$, so that $x_k = P_k + \sqrt{n}$. Hence
$$a_k = [x_k] = P_k + [\sqrt{n}] = P_k + a_0.$$
Also
$$Q_{k+1} = n - P_{k+1}^2 = n - (a_k - P_k)^2,$$
so that
$$Q_{k+1} = n - a_0^2,$$
which is precisely Q_1. Since
$$P_{k+1} = a_k Q_k - P_k = a_k - P_k = a_0,$$
which is precisely P_1, we have proved part of the following.

2.5 Proposition *Suppose that $Q_k = 1$. Then the pair P_{k+1}, Q_{k+1} is a repeat of the pair P_1, Q_1, and hence the calculation of Ps, Qs, x s and a s starts to repeat: $a_{k+1} = a_1$, $a_{k+2} = a_2$, etc. Conversely, suppose that $P_{k+1} = P_1$ and $Q_{k+1} = Q_1$ for some $k > 0$. Then $Q_k = 1$.*

Proof of 'conversely'.   Since $Q_1 = n - a_0^2$ and $P_1 = a_0$ the hypothesis implies, using the formula $Q_{k+1} = (n - P_{k+1}^2)/Q_k$ of 2.2, that

$$n - a_0^2 = (n - a_0^2)/Q_k,$$

and hence $Q_k = 1$ since n is not a perfect square.   □

Notice that this does not quite prove that, if recurrence takes place, then $Q_k = 1$: possibly recurrence takes place from some point other than the second term a_1. Also it does not prove that recurrence takes place at all. Nevertheless both these things are true, and are proved in Section 3 below.

We use the criterion $Q_k = 1$ to signal the onset of recurrence in the following program.

2.6 Program for continued fraction of $\sqrt{n}$

```
PROGRAM P10_2_6;
VAR
n, capP, capQ, rootn, x, a:   extended;
k, flag:   integer;
```

The curious names capP and capQ are used since generally PASCAL *does not distinguish between lower and upper case letters. We need p and q for the convergents later. The flag is used to signal $Q = 1$ but we want to go round one more time and calculate a so the flag has three possible values. Maybe you can find a more ingenious method.*

```
BEGIN
  writeln('input n');
  readln(n);
  rootn:= INT(SQRT(n));
  capP:= 0; capQ:= 1;
  k:= 0; flag:= 0;
  WHILE (flag<2) DO
    BEGIN
      IF (flag = 1) THEN flag = 2;
      x:= (capP + rootn)/capQ;
      a:= INT(x);
      writeln(k,',', capP:0:0,',', capQ:0:0,',', a:0:0);
```

We need to write capP, capQ and a (and x if desired) at this stage since these are all associated to suffix k. Now we go on to calculate capP and capQ associated to suffix $k + 1$. You may want to tabulate the numbers better by inserting spaces between the printed-out values.

```
      capP:= a*capQ - capP;
      capQ:= (n - capP*capP)/capQ;
      k:= k+1;
      IF (capQ = 1) THEN flag:= 1;
   END;
END.
```

2.7 Computing exercises

(a) Find the continued fraction expansions of $\sqrt{n}$ for $n = 11$, 18, 27 and 38. Formulate a general result for numbers of the form $n = d^2 + 2$, and prove it.

(b) Do the same as (a) for $n = 7$, 14, 23 and 34 and formulate and prove a general result.

(c) Find the continued fraction expansion of $\sqrt{1141}$. What is the period length here? Verify that, removing a_0 and the last a_k of the period, what remains reads the same backwards as forwards. (This remarkable observation holds for all n; see for example [K.H. Rosen (1988), §10.4].)

(d) For $n = 13\,290\,059$ verify that $Q_{25} = Q_{29}$. (For this you do not really want to go on to the end of the period, which is very long. Change the program so that it stops at $k = 30$ say.)

(e) (You will need this program to solve Pell's equation in Section 4 below.) Change program P10_2_6 so that it stops when k reaches a predetermined value (which should be input with n), and also prints out the numerator p_k and denominator q_k of each convergent (see 1.2 above for the method of calculating convergents). Also print out the numerical value of p_k/q_k. (If space is getting limited across the screen then you can drop the printing of P_k and Q_k, but of course your program must continue to calculate these since they are needed in the calculation of the a s.) Choose some values of n and verify that the values of p_k/q_k approach $\sqrt{n}$, and are alternately above and below $\sqrt{n}$.

There is one more result we shall prove here about the continued fraction expansion of $\sqrt{n}$. This result is in fact the key (or one of the keys!) to the factorization method described below, Section 5 and also in the solution of Pell's equation in Section 4.

2.8 Proposition *In the usual notation for the continued fraction expansion of $\sqrt{n}$, where n is not a perfect square, we have, for $k \geq 0$,*

$$p_k^2 - nq_k^2 = (-1)^{k+1}Q_{k+1}.$$

Proof We use the same trick as in Section 1 above, where we replaced part of a continued fraction by a complete quotient. Explicitly, we use

$$\sqrt{n} = [a_0, a_1, a_2, \ldots, a_k, \ldots] = [a_0, a_1, a_2, \ldots, a_k, x_{k+1}].$$

(You may be happier verifying the result of the proposition directly for $k = 0$ and assuming $k \geq 1$ below.) Now the convergents p_r/q_r for the second of these coincide with those for the first, for $r \leq k$, since the partial quotients a_i are indentical. Furthermore the convergent p_{k+1}/q_{k+1} for the second is actually equal to $\sqrt{n}$. Using the recurrence formula in 1.2 we obtain

$$\sqrt{n} = \frac{x_{k+1}p_k + p_{k-1}}{x_{k+1}q_k + q_{k-1}}.$$

Now $x_{k+1} = (P_{k+1}+\sqrt{n})/Q_{k+1}$ from the general formulas (2.2 above). Substituting for x_{k+1} and rearranging give the following equation:

$$nq_k - P_{k+1}p_k - Q_{k+1}p_{k-1} = (p_k - P_{k+1}q_k - Q_{k+1}q_{k-1})\sqrt{n}.$$

This has the form $A = B\sqrt{n}$, where A and B are integers. If $B \neq 0$ then this gives $\sqrt{n} = A/B$, which is impossible since $\sqrt{n}$ is irrational. Hence $B = 0$ and so $A = 0$ too. Thus $Bp_k - Aq_k = 0$ and this gives

$$p_k^2 - nq_k^2 = Q_{k+1}(p_k q_{k-1} - p_{k-1}q_k) = Q_{k+1}(-1)^{k+1},$$

using 1.3(b) for the last step (but the power $k+1$ looks more harmonious here!).   □

3. Proofs of some properties of continued fractions

In this section we give a brief account of two matters: convergence of continued fractions (do the convergents converge and if so to what?) and recurrence of the continued fraction for $\sqrt{n}$. Some details are omitted so this section can be regarded as an additional source of exercises.

First, convergence. Suppose in fact that we start with a real positive number x_0 and go through the procedure of 1.1 above to obtain a continued fraction $[a_0, a_1, a_2, \ldots]$ with complete quotients x_k and convergents $c_k = p_k/q_k$ as usual.

3.1 Theorem   *With the above notation, $p_k/q_k \to x_0$ as $k \to \infty$.*

Note. If x_0 is rational then the continued fraction *terminates* with the exact value of x_0.

Proof   The same argument as that used in the proof of 2.8 shows that, for any $k \geq 1$,

$$x_0 = \frac{x_{k+1}p_k + p_{k-1}}{x_{k+1}q_k + q_{k-1}}.$$

It is easy to check that this fraction lies between p_k/q_k and p_{k-1}/q_{k-1}

(this depends only on everything in sight being positive). Also, using 1.3(b), we find

$$\frac{p_k}{q_k} - \frac{p_{k-1}}{q_{k-1}} = \frac{(-1)^k}{q_k q_{k-1}}, \text{ so that } \left| x_0 - \frac{p_k}{q_k} \right| < \frac{1}{q_k q_{k-1}}.$$

However, from the recurrence formula for the qs (see 1.2), and the fact that $a_i \geq 1$ for all $i > 0$, it is clear that the qs form an increasing sequence of positive integers. Thus the above inequality shows that the difference between x_0 and the kth convergent tends to 0 as $k \to \infty$, which is what we wanted to prove. □

3.2 Remark   It can also be shown, equally easily, that given any continued fraction $[a_0, a_1, a_2, \ldots]$ where the a_i are integers ≥ 1 for all $i \geq 0$ (or else stop after a finite number of terms), the convergents converge to a limit x_0 and that $[a_0, a_1, a_2, \ldots]$ is then the continued fraction for x_0 as constructed in 1.1 above.

Now we turn to the form of the continued fraction for $\sqrt{n}$, in particular making precise the signal for recurrence which was used in Program P10_2_6. When studying the continued fraction for $\sqrt{n}$ we considered many numbers of the form $x = \alpha + \beta\sqrt{n}$, where α and β are rational numbers; for example, all the complete convergents x_k have this form. For such a number (which is said to belong to the *field* $Q[\sqrt{n}]$ *generated by* 1 *and* $\sqrt{n}$) we can consider the *conjugate*, defined by $\overline{x} = \alpha - \beta\sqrt{n}$.

Suppose that x is a number of the above form which is not an integer, and let $[x] = a$, so that $0 < x - a < 1$. Define z by $x = a + \frac{1}{z}$ (that is, $z = 1/(x - a)$). An easy calculation shows that $\overline{z} = 1/(\overline{x} - a)$.

3.3 Lemma   *With the above notation we have:*

   If $1 < x$ *and* $-1 < \overline{x} < 0$, *then* $1 < z$ *and* $-1 < \overline{z} < 0$.

Proof   This is an exercise in the use of [ ]. Assume that $1 < x$ and $-1 < \overline{x} < 0$. The proof that $1 < z$ is left as an exercise. For the other part, note that

$$-1 < \overline{z} < 0 \Leftrightarrow -1 < \frac{1}{\overline{x} - [x]} < 0 \Leftrightarrow \overline{x} < [x] \text{ and}$$

$$-\overline{x} + [x] > 1 \Leftrightarrow \overline{x} < [x] - 1.$$

But $1 < x$, so $1 \leq [x]$, and $\overline{x} < 0$ so $\overline{x}+1 < 1 \leq [x]$. Hence $\overline{x} < [x]-1$. □

We now want to deduce results about the continued fraction of $\sqrt{n}$, but it is better to start with $x_0 = [\sqrt{n}] + \sqrt{n}$. The only difference between the continued fraction of this and of $\sqrt{n}$ occurs, naturally, in the first term, which has $[\sqrt{n}]$ added to it. Thus if $\sqrt{n}$ has continued fraction

$$\sqrt{n} = [a_0, a_1, a_2, a_3, \ldots]$$

then x_0 has continued fraction expansion

$$x_0 = [b_0, b_1, b_2, b_3, \ldots],$$

where $b_0 = 2a_0$, $b_k = a_k$ for $k \geq 1$. We shall use the bs in what follows to avoid having to make an exception for $k = 0$. The x_k for $k \geq 1$ are the same for both continued fractions.

Note that $\overline{x_0} = [\sqrt{n}] - \sqrt{n}$, and that $x_0 > 1$, $-1 < \overline{x_0} < 0$ (these would not hold if we had started with $\sqrt{n}$ instead of this x_0). It follows from Lemma 3.3 that all subsequent xs have the same property:

3.4 $x_k > 1$ and $-1 < \overline{x_k} < 0$, for all $k \geq 0$.

Since the x_k for $k \geq 1$ are the same for $\sqrt{n}$ and for $[\sqrt{n}] + \sqrt{n}$, the integers P_k and Q_k obtained from $x_k = (P_k + \sqrt{n})/Q_k$ will, for $k \geq 1$, be those encountered in Section 2 for the continued fraction expansion of $\sqrt{n}$. We can now prove:

3.5 Theorem   *The P_k and Q_k associated with the continued fraction expansion of $\sqrt{n}$ satisfy $0 < P_k < \sqrt{n}$ and $0 < Q_k < 2\sqrt{n}$, for all $k \geq 0$.*

Proof   Since $P_0 = 0$ and $Q_0 = 1$ there is no difficulty with $k = 0$; for $k \geq 1$ we use the above results. Thus

$$x_k = \frac{P_k + \sqrt{n}}{Q_k}, \quad \overline{x_k} = \frac{P_k - \sqrt{n}}{Q_k} \quad (k \geq 1).$$

If $Q_k < 0$ then using 3.4 these give $Q_k > P_k + \sqrt{n}$ and $-Q_k > P_k - \sqrt{n} > 0$, which imply that $\sqrt{n} < P_k < -\sqrt{n}$, a contradiction. Since in 2.2 above we found $Q_k \neq 0$, we can deduce $Q_k > 0$.

Applying 3.4 again, we find $| P_k - \sqrt{n} | < Q_k$ and $| P_k + \sqrt{n} | > Q_k$, which implies $P_k > 0$, and $\overline{x_k} < 0$ now gives $0 < P_k < \sqrt{n}$. Finally $P_k + \sqrt{n} > Q_k$ gives $Q_k < 2\sqrt{n}$. $\square$

Since the Ps and Qs are restricted in size as in 3.5, so that there are only a finite number of possible pairs (P_k, Q_k), we deduce:

3.6 Corollary   *There must exist r and s with $0 \leq r < s$ and $P_r = P_s$, $Q_r = Q_s$, and hence $x_r = x_s$.* $\square$

The final result which we prove here is that the continued fraction for $x_0 = [\sqrt{n}] + \sqrt{n}$ recurs from the beginning:

3.7 Theorem   *The continued fraction for $[\sqrt{n}] + \sqrt{n}$ has the shape*

$$[\sqrt{n}] + \sqrt{n} = [2a_0, a_1, a_2, \ldots, a_{s-1}, 2a_0, a_1, a_2, \ldots, a_{s-1}, 2a_0, a_1, a_2, \ldots],$$

so that the continued fraction for $\sqrt{n}$ has the shape

$$\sqrt{n} = [a_0, \overline{a_1, a_2, \ldots, a_{s-1}, 2a_0}].$$

Furthermore $Q_s = 1$ and $Q_k > 1$ for $0 < k < s$.

The **Proof** is given here as a series of exercises:

(a) Let $y_k = -1/\overline{x_k}$ in the notation above. Check that $y_k = b_{k-1} + 1/y_{k-1}$, for $k \geq 1$. (Recall $b_k = a_k$ for $k \geq 1$, $b_0 = 2a_0$. Use $P_k = b_{k-1}Q_{k-1} - P_{k-1}$ for $k \geq 2$.)

(b) Use 3.4 to show that $y_k > 1$ for all $k \geq 0$ and hence that $[y_k] = b_{k-1}$ for $k \geq 1$.

(c) Let now $x_r = x_s$ where $0 \leq r < s$ and r is as small as possible (see 3.6). Deduce that $y_r = y_s$ and that, if $r > 0$, then $b_{r-1} = b_{s-1}$. Now use $x_{r-1} = b_{r-1} + 1/x_r$ to deduce $x_{r-1} = x_{s-1}$, contradicting the minimality of r. Deduce that indeed $r = 0$, so that $x_0 = x_s, b_0 = b_s$, where we may suppose this is the first repetition.

(d) Deduce the results of 3.7, bearing in mind 2.5 above.

3.8 Project Given a periodic continued fraction which does not necessarily repeat from the beginning,

$$[a_0, a_1, a_2, \ldots, a_{s-1}, \overline{a_s, a_{s+1}, \ldots, a_{s+k}}],$$

it can be shown that this equals a 'quadratic irrational' $(P_0 + \sqrt{n})/Q_0$ for some integers P_0, Q_0 with $Q_0 \neq 0$. See for example [K.H. Rosen (1988), Section 10.4]. Read up the theory and write a program to determine P_0, n and Q_0 from s, k and $a_0, \ldots, a_{s+k}$.

4. Pell's equation

Before going on to the factoring algorithm, we pause here to use Proposition 2.8 above to solve a famous number-theoretic equation: for a given integer $n \geq 2$

$$x^2 - ny^2 = 1$$

is called *Pell's equation* for the integers $x, y > 0$.

The equation was considered in several special cases by W. Brouncker (1657), by Fermat (see for example 4.3(b) below) and indeed much earlier, in the twelfth century, by Indian mathematicians. It was wrongly attributed to John Pell (1611–1685) by Euler; the name has stuck ever since.

If n is a perfect square then in effect we are solving $x^2 - z^2 = 1$, that is $(x - z)(x + z) = 1$, which is easily seen to have no solutions > 0. Thus we assume n is a nonsquare. For example, $x^2 - 3y^2 = 1$ has solutions

$(x, y) = (2, 1), (7, 4), (26, 15), (97, 56), \ldots$, while the smallest solution of $x^2 - 109y^2 = 1$ is $(x, y) = (158\,070\,671\,986\,249, 15\,140\,424\,455\,100)$.

4.1 Exercises

(a) Suppose that $x^2 - ny^2 = -1$. Show that
$$(x^2 + ny^2)^2 - n(2xy)^2 = 1.$$

(b) Verify directly that $8\,890\,182^2 - 109(851\,525)^2 = -1$. Hence, using (a), find the solution to $x^2 - 109y^2 = 1$ given above.

(c) As a less extreme example, find a very small solution to $x^2 - 10y^2 = -1$ and hence a solution to $x^2 - 10y^2 = 1$ (but with $x > 0$ and $y > 0$ of course!). Generalize to numbers n of the form $n = m^2 + 1$.

From Proposition 2.8 above we deduce (a) of the following.

4.2 Proposition (a) *Suppose that, in the usual notation for the continued fraction of $\sqrt{n}$, we have $Q_s = 1$ (compare 3.7 above). Then $p_{s-1}^2 - nq_{s-1}^2 = (-1)^s$, so that (p_{s-1}, q_{s-1}) is a solution to Pell's equation if s is even. If s is odd, then, using 4.1(a), $(p_{s-1}^2 + nq_{s-1}^2, 2p_{s-1}q_{s-1})$ is a solution.*

(b) *If s is the smallest number > 0 giving $Q_s = 1$, then $Q_{ms} = 1$ for $m = 1, 2, 3, \ldots$. Thus, when s is even, $(-1)^{ms} = 1$ for all m, and (p_{ms-1}, q_{ms-1}) is a solution of Pell's equation. When s is odd, the even multiples $2ms$ give $(-1)^{2ms}Q_{2ms} = 1$ and (p_{2ms-1}, q_{2ms-1}) is a solution to Pell's equation.*

Proof of (b) Referring back to Theorem 3.7, the period of the continued fraction is s and it follows that $Q_{ms} = 1$ for all $m \geq 1$. The remainder of the statement now follows.   □

For example, $\sqrt{7} = [2, \overline{1, 1, 1, 4}]$, with even period $s = 4$, so that $Q_4 = 1$ and (p_{4m-1}, q_{4m-1}) is a solution to Pell's equation $x^2 - 7y^2 = 1$. The first few are $(8, 3), (127, 48), (2024, 765)$.

On the other hand, $\sqrt{41} = [6, \overline{2, 2, 12}]$ has odd period $s = 3$, so that (p_{6m-1}, q_{6m-1}) give solutions to Pell's equation (starting with $(2049, 320), (8\,396\,801, 1\,311\,360))$, while (p_{6m+2}, q_{6m+2}) give solutions to $x^2 - 41y^2 = -1$, the first two being $(32, 5)$ and $(131\,168, 20\,485)$.

In fact, with a little more trouble, it can be shown that the solutions to Pell's equation found in this way form the *complete set of solutions*. See for example [K.H. Rosen (1988),§11.4].

4.3 Computing exercises

(a) Use the program in Exercise 2.7(e) above which calculates the continued fraction expansion of $\sqrt{n}$ to find solutions to $x^2 - ny^2 = 1$, for

n taking the values 47, 52, 29. In each case find solutions up to extended precision. (The values of s for these values of n are respectively 4, 6, 5, as you will find from printing out the a s or the Q s, either to look for recurrence or to look for the first 1.)

(b) Find the smallest solution to $x^2 - 61y^2 = 1$ (that is the smallest solution given by the above method; in fact this *is* the smallest solution). Fermat, in a letter to his friend Frénicle in 1657, suggested that he solve this equation 'pour ne vous donner trop de peine'. ...

5. The continued fraction factoring method

The method we shall give was first described by D.H. Lehmer and R.E. Powers in 1931. It was subsequently developed and implemented on a computer by M.A. Morrison and J. Brillhart, who successfully factored the gigantic Fermat number $F_7 = 2^{128} + 1$ as

$$59\,649\,589\,127\,497\,217 \times 5\,704\,689\,200\,685\,129\,054\,721$$

on the morning of 13 September, 1970. (They had spent more than the previous night working on the problem!) See [Morrison and Brillhart (1975)]. Both these factors are actually prime, so this is a complete factorization. A generalization of this method was used in 1990 to factorize the even more gigantic $F_9 = 2^{512} + 1$, which has 155 digits. This took two months on 1000 computers around the world. See [Cipra (1990)].

Note that we *seek a (proper) factor of n*, not necessarily to factorize n completely. Once a proper factor x has been found, that factor and the quotient n/x can be further subjected to the method, if a complete factorization is wanted. We speak of 'factoring n' to mean 'finding a proper factor'.

The congruence

$$p_k^2 \equiv (-1)^{k+1}Q_{k+1} = Q_{k+1}^* \qquad (\text{mod } n)$$

follows at once from 2.8 above; Q_i^* stands for $(-1)^i Q_i$ here and in what follows. What we do is to look for products of the Q^*s which are perfect squares; the purpose of this will be explained below. Note the important fact (see 3.5 above) that all the $|\, Q_i^* \,|$ are $< 2\sqrt{n}$; consequently we can expect that factorizing the Q^*s in order to find such products will be a less arduous task than finding a factor of n !

5.1 Exercise Suppose that the prime p divides Q_{k+1}, for some $k \geq 0$. Use 2.8 above to show that $p_k^2 \equiv nq_k^2 \pmod{p}$. Use the coprimality of p_k and q_k (see 1.3(b) above) to show that p cannot divide q_k, and hence that there exists r with $rq_k \equiv 1 \pmod{p}$. Deduce that there exists x with $x^2 \equiv n \pmod{p}$. We say that n is a *quadratic residue* mod p, meaning that it is congruent to a square mod p. This observation does in fact cut down considerably on the primes which might be factors of the Qs. We shall return to this topic in Chapter 11.

The process of looking for products of the Q^*s which are squares (this does mean *positive* squares: remember Q_i^* has a sign attached) can be automated, and this is done in the full computer implementation of the factoring method. But here we shall do this part by hand.

5.2 Basis of the factoring method Suppose that we have found a product

$$Q_{k_1+1}^* Q_{k_2+1}^* \cdots Q_{k_m+1}^* = z^2.$$

Note that this is =, not ≡ : it is easier to look for products which *equal* squares than products which are merely *congruent to squares.* Then we have

$$p_{k_1}^2 p_{k_2}^2 \cdots p_{k_m}^2 \equiv z^2 \pmod{n}.$$

In this expression we may reduce each p_i mod n: in practice we replace each p_i by its residue of least *absolute value*, that is by a number x with $-n/2 < x < n/2$.

Thus we have found two numbers whose squares are congruent mod n. We consider the greatest common divisor

$$(p_{k_1} p_{k_2} \cdots p_{k_m} - z, n)$$

which is of course a factor of n. With luck (compare Chapter 3,1.5) it will be a *proper* factor of n, that is neither 1 nor n. Occasionally it turns out that this does not give a proper factor but

$$(p_{k_1} p_{k_2} \cdots p_{k_m} + z, n)$$

does (see 5.6(d) below). So it is worth working this out too (we do this in the program P10_5_5).

5.3 Example $n = 1711$. We find the following table (produced by a program we shall give later; of course * means product):

$k+1$	$p_k \pmod{n}$	Q_{k+1}^*		factorization
1	41	-30	=	$-2*3*5$
2	83	45	=	$3*3*5$
3	124	-23	=	-23
4	331	57	=	$3*19$
5	455	-6	=	$-2*3$
6	-598	5	=	5
7	-558	-38	=	$-2*19$

8	−3	9	=	3*3
9	−582	−54	=	−2*3*3*3
10	−585	25	=	5*5
11	−41	−30	=	−2*3*5
12	−667	29	=	29
13	336	−30	=	−2*3*5
14	5	25	=	5*5
15	346	−54	=	−2*3*3*3
16	351	9	=	3*3
17	−268	−38	=	−2*19
18	−185	5	=	5
19	194	−6	=	−2*3
20	626	57	=	3*19
21	820	−23	=	−23
22	555	45	=	3*3*5
23	−336	−30	=	−2*3*5
24	−117	1	=	1	Note Q = 1:recurrence
25	336	−30	=	−2*3*5
26	555	45	=	3*3*5
27	−820	−23	=	−23
28	626	57	=	3*19
29	−194	−6	=	−2*3
30	−185	5	=	5
31	268	−38	=	−2*19
32	351	9	=	3*3
33	−346	−54	=	−2*3*3*3
34	5	25	=	5*5
35	−336	−30	=	−2*3*5
36	−667	29	=	29
37	41	−30	=	−2*3*5
38	−585	25	=	5*5
39	582	−54	=	−2*3*3*3
40	−3	9	=	3*3

Notice that the Q^*s recur after row 24 (since 24 is even; otherwise they would recur with a minus sign). However the ps, naturally, don't recur! In fact rows 1 and 25 (we refer here and later to the rows labelled 1 and 25; of course these are values of $k + 1$, not k) give

$$Q_1^* Q_5^* = (2 \cdot 3 \cdot 5)^2 \text{ so } p_0^2 p_{24}^2 \equiv 41^2 \cdot 336^2 \equiv (2 \cdot 3 \cdot 5)^2 \pmod{1711}.$$

This says that $13\,776^2 \equiv 30^2 \pmod{1711}$, and as it happens $(13\,776 -$

30, 1711) is the *proper* factor 29 of 1711, so the method has produced a proper factor. It is not generally feasible, with even moderately sized n, to calculate as far as recurrence.

Another pair of rows which works here is rows 2 and 18:

$$(83 \cdot 185)^2 \equiv (5 \cdot 3)^2, \text{ and } (83 \cdot 185 - 15, 1711) = 29.$$

On the other hand, rows 2 and 6 don't work:

$$(83 \cdot 598)^2 \equiv (3 \cdot 5)^2, \text{ but } (83 \cdot 598 - 15, 1711) = 1711$$

(and it's no good using $(83 \cdot 598 + 15, 1711)$ either, since that's 1).

5.4 Exercises

(a) Show that rows 2 and 26 give the factor 59; that row 8 by itself is no use, and neither is row 14, whereas row 10 by itself does give the factor 29. Show that rows 4, 7 and 9 give the factor 29.

(b) Find some other combinations of rows which give a proper factor of 1711.

We give below a program for producing a table such as that above for $n = 1711$. It is simply a matter of incorporating a program for finding the continued fraction expansion of $\sqrt{n}$ (or the ps and Q_ks of such an expansion) with a factorizing program to break the Q^* up into factors. We shall give some more examples shortly where the program can be used; it is worth reiterating here that this section is designed only to explain the working of the continued fraction algorithm, not to present realistically large numbers to factor. The finding of products of Q^* which are perfect squares must of course be automated for very large n. Here we invite you to find your own combinations and then to try them out.

In the program below, you are first asked for n (the number to be factored) and the number ($maxk$) of rows of the table to be calculated. You are then asked for *maxprime*, which is the largest prime which is to be used in the attempted complete factorizations of the Q^*s. We look for such numbers which factorize into relatively small primes, and the factorization is accomplished by program P2_1_13, modified along the lines of 1.15 in Chapter 2, written as the procedure Factors. When the number maxprime is exceeded in a factorization, the program prints out ***, meaning: ignore this row of the table. The factorization, when successful, is printed out thus: 60 = 2*2*3*5*; with a modicum of ingenuity you can of course eliminate the final *, as was done in the printout for 1711 above.

At the end of the program you are invited to say whether you want to attempt to find a factor of n; if you press 'Y' or 'y', for 'yes', then you are asked for the number of rows which are used to make the product of the Q^*s a perfect square, and then for the row numbers. The gcd is calculated by means of program P1_3_5, written as the procedure Euclid.

You could improve the program by using Head's algorithm for working out the products of the ps mod n, but the examples below do not assume that this has been done. WARNING: If you multiply very many Q^*s together before taking the square root then the product will exceed double precision. This is obviously a limitation of this program; it would be better to store the factorizations of the Q^*s and to calculate the square root of the product by taking just half the factors of the product. Maybe you would like to try this!

5.5 Program for finding a factor of n by the continued fraction method

```
PROGRAM P10_5_5;
CONST
arraysize = 100;    (* If you want to go beyond the *)
(* 100th row then increase the array size !*)
VAR
n, rootn, capP, capQ, Qstar, p, pp, ppp, p1:   extended;
maxprime, a, b, lhs, rhs, diff, sum, gcd, number :
extended; p_array, Qstar_array:   ARRAY[1..arraysize]
OF extended; maxk, i, k, rownum, terms :   integer;
yorn:   char;    (* A yes/no answer *)
f:   text;    (* The file of primes *)

  PROCEDURE Factors(number, maxprime:   extended);
  VAR prime:extended;

  BEGIN
    IF (number < 0) THEN
      BEGIN
        write(' - ');
        number:= - number;
      END;
    reset(f);
    WHILE number>1 DO
      BEGIN
        read(f,prime);
        IF (prime <= maxprime) THEN
          BEGIN
            WHILE (number/prime = INT(number/prime)) DO
              BEGIN
```

```
                 write(prime:0:0,'*');
                 number:=number/prime;
               END;
           END
         ELSE
           BEGIN
             write('***');    (* This indicates that *)
             (* the prime exceeds maxprime *)
             number:=1;
           END;
       END;
     writeln;
     close(f);
   END;   (* of Factors *)

   PROCEDURE Euclid(a, b:  extended; VAR gcd:  extended);
   VAR r :  extended;

   BEGIN
     WHILE (b>0) DO
       BEGIN
         r:= a - b*INT(a/b);
         a:= b;
         b:= r;
       END;
     gcd:= a;
   END;   (* Of Euclid *)

 BEGIN   (* main program *)
   Assign (f,'primes.dat');
   writeln ('Type the value of n followed by the
     maximum k+1');
   readln (n,maxk);
   writeln ('Type the value of a bound for the primes');
   writeln ('used  in factorizing by trial division');
   readln (maxprime);
   writeln (' n = ', n:0:0);
   writeln ('k + 1      p[k] mod n      Q[k+1]* ');
   rootn:=INT(SQRT(n));
   capP:=0;
   capQ:=1;
   pp:=1;
```

```
ppp:=0;
i:=1;
WHILE (i<=maxk) DO
  BEGIN
    a:=INT((capP + rootn)/capQ);
    p:=a*pp + ppp;
    p:=p - INT(p/n)*n;
    capP:=a*capQ - capP;
    capQ:=(n - capP*capP)/capQ;
    IF (i<>Int(i/2)*2) THEN
      Qstar:= - capQ
    ELSE
      Qstar:=capQ;
    IF (p>n/2) THEN
      p1:=p - n
    ELSE
      p1:=p;
    (* So p1 is the smallest residue in absolute *)
    (* value of p *)
    ppp:=pp;
    pp:=p;
    write (i:4, p1:11:0, Qstar:11:0, ' = ');
    Factors (Qstar,maxprime);
    p_array[i]:=ABS(p1);
    (* We don't need to bother with the sign of p *)
    Qstar_array[i]:=Qstar;
    i:= i + 1;
  END;    (* of WHILE i<=k *)

(* The rest of the program tries the factoring *)
(* method on n *)

writeln ('Do you wish to try a factor?  (y/n)');
readln (yorn);    (* meaning "yes or no" ! *)
WHILE (yorn='y') OR (yorn='Y') DO
  BEGIN
    lhs:=1;
    rhs:=1;
    writeln ('How many terms to make a square?');
    readln (terms);

    FOR i:=1 TO terms DO
```

```
        BEGIN
          writeln ('What is the number of the row?');
          readln (rownum);
          lhs:=lhs*p_array[rownum];
          lhs:=lhs - n*INT(lhs/n);
          rhs:=rhs*Qstar_array[rownum];
        END;

      rhs:=sqrt(rhs);

      (* At this point, lhs is the product, mod n, of *)
      (* the p's and rhs is the square root of the *)
      (* product of the Q*'s. There is a danger of *)
      (* loss of accuracy if the product of Q*'s is *)
      (* too big. Next the difference, diff, will be *)
      (* computed. Its gcd with n is a factor *)
      (* c(hopefully proper) of n. We also compute *)
      (* the sum, called sum, since this sometimes *)
      (* works when diff doesn't *)

      writeln ('lhs = ', lhs:0:0, ' rhs = ', rhs:0:0);
      diff:=lhs - rhs;
      Euclid(diff, n, gcd);
      writeln ('A factor of ', n:0:0, ' is ', gcd:0:0);
      sum:=lhs + rhs;
      Euclid(sum,n,gcd);
      writeln ('A factor of ', n:0:0, ' is ', gcd:0:0);
      writeln ('Do you wish to try another factor?
      (y/n)');
      readln (yorn);
    END    (* of yorn is 'yes' *)
  END    (* of main program *)
```

5.6 Computing exercises

(a) Check that the program gives the same table as is printed out above for $n = 1711$, and the same factorings that we had before.

(b) Try $n = 13\,290\,059$ with $maxk = 40$ and $maxprime = 600$. This can be factored using rows 18 and 36. Find other combinations of rows which give a proper factor. (You will find it a great help if you can print out the data from the factorizing of the Q^*s, and look for products giving squares using the printed version.)

(c) There is a square among the Q^*s for $n = 104\,838\,947$ before the

60th row. Find this and use it to find a proper factor of n. Also find other combinations of rows yielding proper factors.

(d) There is a product of five rows (all with row numbers < 50) which factorizes the number $n = 12\,007\,001$; find these or another combination which factorizes n. (This is an example where using the *sum* of the product of ps and the square root of the product of Q^*s gives a proper factor while the *difference* doesn't.) This example is also about at the limit of accuracy for calculating the square root of the product of Q^*s by first computing the product of Q^*s and then taking the square root. See the WARNING just before the program.

(e) $n = 1\,633\,961\,543$ can be factored using the 354th row, which has Q^*_{354} a perfect square! (If you check this then don't forget to increase the constant '*arraysize*' at the beginning of the program.) Maybe you can find earlier rows which do the trick.

5.7 Use of a multiplier At the beginning of this section we said that one of the triumphs of the continued fraction method was the factorization of $2^{128} + 1$ in 1970. However, this number has the form $d^2 + 1$, and referring to 2.4(b) above you will see that the continued fraction expansion of a number of this form recurs very quickly. This implies that there are very few distinct Qs and so not much hope of finding a useful product which is a square. (What *are* the Q^*s in this case?) The remedy lies in the use of a *multiplier*: instead of calculating the continued fraction expansion of $\sqrt{n}$ we calculate that of $\sqrt{\lambda n}$ for some 'suitable' positive integer λ. Morrison and Brillhart used $\lambda = 257$ in their factorization of F_7.

Note that while the continued fraction is for $\sqrt{\lambda n}$, all reductions are still done modulo n. Try amending program P10_5_5 to cope with this slightly altered situation. As a very simple example try $n = 4097$ with the original P10_5_5; you will agree that there is not much hope from that; now try a multiplier of $\lambda = 3$. There is a square among the Q^*s quite early in the expansion and this successfully finds a factor.

A slightly more robust example is provided by $n = 2^{29} - 1 = 536\,870\,911$ with multiplier $\lambda = 3$. The second perfect square occurring among the Q^*s gives a nontrivial factor.

Perhaps you can find a multiplier which enables the Fermat number $F_5 = 2^{32} + 1 = 4\,294\,967\,297$ to be factored.

11

Quadratic residues

1. Definitions and examples

We saw in Chapter 10, 5.1, that the primes p which are factors of the numbers Q^* occurring in the continued fraction expansion of $\sqrt{n}$ have a special property: for each such p there is an x with $x^2 \equiv n \pmod{p}$.

1.1 Definition *Let p be an odd prime and let n be such that $p \nmid n$. If $n \equiv x^2 \pmod{p}$ for some x, that is n is a square modulo p, then we say that n is a* quadratic residue modulo p, *and write nRp. If on the other hand no such x exists then we say that n is a* quadratic nonresidue modulo p *and write nNp. We shall also use the* Legendre symbol $\left(\frac{n}{p}\right)$, *introduced by A.M. Legendre in 1798. It is defined when $p \nmid n$: $\left(\frac{n}{p}\right) = +1$ if nRp and $\left(\frac{n}{p}\right) = -1$ if nNp.*

Of course, $\left(\frac{n^2}{p}\right) = 1$ for any n, provided $p \nmid n$; in particular $\left(\frac{1}{p}\right) = 1$ for any p.

The case $p = 2$ is omitted because it is uninteresting (every odd n is $\equiv 1^2 \pmod{2}$). For any particular p we can find the n for which nRp by squaring the numbers $x = 1, 2, \ldots, p-1$ in turn; the values $n \equiv x^2 \pmod{p}$ are precisely those required. For example with $p = 5$ the squares are $1, 4, 4, 1$ so that $1R5$, $4R5$, while $2N5$, $3N5$. It will turn out to be much easier than this, however, to determine whether nRp for a given n and p.

1.2 Computing exercise Write a program to calculate the squares of the numbers $1, 2, \ldots, p-1 \pmod{p}$ and hence verify that for every p you try, there are the same number, $(p-1)/2$, of residues and nonresidues mod p.

The observation of 1.2 follows, in fact, from 2.5 in Chapter 8, which we recall here since it will play an important role in what follows.

1.3 Proposition: Euler's criterion for quadratic residues *Let g be a primitive root* mod p *where p is an odd prime, and let $p \nmid n$, so that $n \equiv g^k \pmod{p}$ for some k. Then*

$$k \text{ is even } \Leftrightarrow n^{(p-1)/2} \equiv 1 \pmod{p} \Leftrightarrow nRp.$$
$$k \text{ is odd } \Leftrightarrow n^{(p-1)/2} \equiv -1 \pmod{p} \Leftrightarrow nNp.$$

1.4 Proposition: basic properties of the Legendre symbol

(a) *Alternative statement of Euler's criterion:* $\left(\frac{n}{p}\right) \equiv n^{(p-1)/2}$ *(mod p) (here p is prime and $p \nmid n$).*

(b) *Provided p is prime, $p \nmid n$ and $p \nmid m$,* $\left(\frac{n}{p}\right)\left(\frac{m}{p}\right) = \left(\frac{nm}{p}\right)$*, that is,*

$$nmRp \Leftrightarrow (nRp \text{ and } mRp) \text{ or } (nNp \text{ and } mNp).$$

(c) *Provided p is prime and $p \nmid n$,* $\left(\frac{n}{p}\right) = \left(\frac{n+kp}{p}\right)$ *for any integer k.*

Proof (a) immediate from 1.3 and the definition in 1.1, while (b) follows immediately from (a). Finally (c) is clear from the definition of quadratic residue. □

The result in 1.4(b) is one overwhelming justification for using the Legendre symbol $\left(\frac{n}{p}\right)$. Note that it would not be very easy to prove (b) directly from the definitions of nRp and nNp.

1.5 Examples

(a) Let us take $n = -1$. Then $\left(\frac{-1}{p}\right) = 1$, that is $-1Rp$, if and only if $(p-1)/2$ is even, that is $p \equiv 1 \pmod{4}$. This is our first example of fixing n and asking for those primes p for which $\left(\frac{n}{p}\right) = 1$. In Chapter 3,2.11(b), it is shown that this implies the existence of integers x and y with $x^2 + y^2 = p$.

(b) As an application of (a), let us prove that *there exist infinitely many primes of the form $4r + 1$* (compare 1.7 in Chapter 1 and 1.19 in Chapter 2). In fact, let n be any positive integer and let $Q = (n!)^2 + 1$. Let p be a prime with $p \mid Q$; clearly no prime $< n$ divides Q, so $p > n$. We claim that p necessarily has the form $4r + 1$, which is enough to prove the result since n is arbitrarily large. In fact, since $p \mid Q$, we have $(n!)^2 \equiv -1 \pmod{p}$, so that $\left(\frac{-1}{p}\right) = 1$, and the result follows from (a).

1.6 Exercises

(a) Suppose that $p \equiv 1 \pmod{4}$. Show that, for any n with $p \nmid n$, $\left(\frac{n}{p}\right) = \left(\frac{-n}{p}\right)$.

(b) Why does the example in Chapter 6,3.6(g), show that the prime $p = 26\,437\,680\,473\,689$ has the unusual property that $\left(\frac{n}{p}\right) = 1$ for all n with $1 \leq n \leq 500$, and hence by (a) for all (nonzero) n with $\mid n \mid \leq 500$?

(c) Prove 1.5(b) above in a different way as follows. Let $p_1, p_2, \ldots, p_k$ be primes of the form $4r + 1$ and consider $Q = (p_1 p_2 \ldots p_k)^2 + 1$. Show that every prime p dividing Q must have the form $4r + 1$, using 1.5(a).

(d) Show that there are infinitely many primes of the form $8r + 5$, as follows. Let $p_1, \ldots, p_k$ be k such primes and consider $Q = (p_1 \ldots p_k)^2 + 4$. Show that $Q \equiv 5 \pmod 8$. Let p be a prime dividing Q (so $p \neq p_i$ for any i and $p \neq 2$); deduce that $\left(\frac{-4}{p}\right) = 1$ and hence that $\left(\frac{-1}{p}\right) = 1$ (see 1.4(b)), so that $p \equiv 1$ or $5 \pmod 8$ from 1.5(a). Why is it impossible for all such p to be $\equiv 1 \pmod 8$?

(e) Why does the observation of 1.2 follow from 1.3?

(f) Let $p = 11$. Find two numbers x and y ($1 \leq x \leq p - 1$, $1 \leq y \leq p - 1$) such that x, y and $x + y$ are distinct and all quadratic residues (mod p). Do such x and y exist for any prime $p < 11$? These x, y form a *chain of quadratic residues*; compare [Gupta (1971)]; see also 1.7(a) below.

(g) Prove **Pepin's test** for the primality of $F_n = 2^{2^n} + 1$ (T. Pepin, 1877): Suppose that $n \geq 2$ and $k \geq 2$. Then the following are equivalent, writing F for F_n:

(i) F is prime and $\left(\frac{k}{F}\right) = -1$; (ii)$k^{(F-1)/2} \equiv -1 \pmod F$.

[Hints. (i) $\Rightarrow$ (ii) is easy. Assuming (ii), use the '$n - 1$ over q test' (3.4 of Chapter 6) to show that F is prime and Euler's criterion to show the rest.] We shall see later (3.5(a)) that $k = 3$ is always a suitable choice.

(h) Why does 1.4(b) imply that the smallest positive quadratic *non-residue* (mod p) is always prime?

1.7 Computing exercises

(a) (following on from (f) above). Taking $p = 631$ and $x_1, \ldots, x_{11}$ to be 1, 4, 45, 94, 261, 310, 344, 387, 393, 394, 456 respectively, verify that these form a 'chain of quadratic residues' (mod 631) in the sense that all the 66 sums $\sum_{k=i}^{j} x_k$ ($1 \leq i \leq j \leq 11$) are distinct quadradic residues (mod 631). (This p is the smallest to have a chain of length 11.)

(b) Use Euler's criterion to check that the prime $p = 30\,059\,924\,764\,123$ has the property that $\left(\frac{n}{p}\right) = -1$ for all *primes* n with $2 \leq n \leq 181$. Compare 1.6(b) above and Chapter 6, 3.6(i). What about composites n? (See 1.4(b).)

(c) Let p and q be distinct odd primes < 100. Calculate $\left(\frac{p}{q}\right)$ and $\left(\frac{q}{p}\right)$ for each possible choice (there are 276 unordered pairs $\{p, q\}$). See if you can spot a rule which says when the product is 1 or -1. (The rule is revealed in Section 3 below.)

1.8 Project (Compare Chapter 4,1.2(e).) Use Euler's criterion to determine those primes $p < 200$ for which -1091 is a quadratic residue mod p. (You will of course want to use the power algorithm, as in P4_2_3. You might want to handle the minus sign separately!) We are interested in the following question (compare [Shanks (1971)]): for a given prime p, which numbers of the form $f(n) = n^6 + 1091$ are divisible by p? In particular, how many of these numbers, for $n = 1, 2, 3, 4, \ldots$, can we prove to be composite? (The first prime turns out to be quite far along the sequence!) Clearly if $p \mid f(n)$ then $\left(\frac{-1091}{p}\right) = 1$, so we need only bother with primes p for which this is so, together with $p = 2$, and $2 \mid f(n)$ if and only if n is odd. The first few p are 3, 5, 7, 11, 13, 19, 23, 41, as you should have discovered! Use Fermat's theorem to show that, if $p \nmid n$, then $p \mid f(n)$, for $p = 2$, 3 and 7. Deduce that, if $f(n)$ is prime, then $42 \mid n$, so that in looking for the first prime of the form $f(n)$ we can write $n = 42m$.

Here are some suggestions for eliminating some more composites among the $f(n)$. Using Euler's criterion, and $1091 \equiv -1 \pmod{13}$, we have $n^6 + 1091$ is divisible by 13 whenever $\left(\frac{n}{13}\right) = 1$. It is not hard to check that this holds for $n \equiv 1, 3, 4, 9, 10$ and 12 (mod 13) (or maybe you did this in 1.2). This implies that if $f(n)$ is prime then $n = 210k \pm 84$ for some k. The first prime number does in fact occur for $k < 20$.

To find when $19 \mid f(n)$ you could use primitive roots (compare 1.2(b) in Chapter 8); 2 is a primitive root mod 19. You should find that the values of n are $\equiv 4$, 6, 9, 10, 13 and 15 (mod 19).

How many of the $f(210k \pm 84)$ with $k < 20$ can you prove to be composite?

2. The quadratic character of 2

Once we have found which primes p have $\left(\frac{2}{p}\right) = 1$ there are many interesting results which follow. In fact we state the answer first, then give some applications, and finally give the proof. (Most of the results below are due originally to Gauss, and were proved by him, sometimes several times, in his classic *Disquisitiones Arithmeticae*. See [Gauss (1986)].)

2.1 Theorem:the quadratic character of 2 $\left(\frac{2}{p}\right) = 1$ *if and only if p is a prime of the form $8k \pm 1$.(Thus $\left(\frac{2}{p}\right) = -1$ for primes of the form $8k \pm 3$.).*

This can be equivalently stated:

$$\left(\tfrac{2}{p}\right) = (-1)^{(p^2-1)/8}.$$

2.2 Examples

(a) Suppose that $p = 4m+3$ and $q = 2p+1$ are both prime. Then $q \equiv -1 \pmod 8$, so that by 2.1 we have $\left(\tfrac{2}{q}\right) = 1$, so that $2^{(q-1)/2} \equiv 1 \pmod q$ by Euler's criterion, which implies $q \mid 2^p - 1$. Hence, the *Mersenne number* $M_p = 2^p - 1$ is *composite*. This, combined with the examples of Chapter 6,3.2 and 3.6(h), shows that the following rather large Mersenne numbers are composite: M_p for $p = 1\,122\,659$ (this M_p has about $338\,000$ digits), M_{2p+1} for the same p, and several other numbers coming from the same Cunningham chain of primes; also M_p for $p = 16\,035\,002\,279$ and $16\,048\,973\,639$. About how many digits do these latter Mersenne numbers have?

(b) We can prove quite easily that *there are infinitely many primes of the form $8r - 1$*. For let $p_1, p_2, \ldots, p_k$ be k such primes and consider $Q = (4p_1...p_k)^2 - 2$. Then $Q \equiv -2 \pmod 8$ and, for any odd prime $p \mid Q$, we have $\left(\tfrac{2}{p}\right) = 1$ so that by 2.1 we have $p \equiv \pm 1 \pmod 8$. However, if all odd primes dividing Q were congruent to 1 (mod 8) then Q itself (being 2 times a product of odd primes) would be congruent to 2 (mod 8), which is not the case. Hence Q has an odd prime divisor of the form $8r - 1$, and since this cannot be any of the p_i the result follows.

(c) We show that *if p and $q = 4p + 1$ are both prime, then 2 is a primitive root for q*. Since $2^{q-1} \equiv 1 \pmod q$, we have $\mathrm{ord}_q\, 2 \mid (q - 1) = 4p$, so that $\mathrm{ord}_q\, 2$, which we want to equal $4p$, must be 1, 2, 4, p, $2p$ or $4p$. The first three are easily eliminated, while $\mathrm{ord}_q\, 2 = p$ or $2p$ implies that $2^{(q-1)/2} \equiv 1 \pmod q$, which by Euler's criterion implies $\left(\tfrac{2}{q}\right) = 1$, so that by 2.1 we have $q \equiv \pm 1 \pmod 8$. But p must be odd so in fact $q \equiv 5 \pmod 8$, a contradiction. Hence the order is $4p$, as required.

2.3 Exercises

(a) Use 2.1, 1.5(a) and 1.4(b) to show that $-2Rp$ if and only if $p \equiv 1$ or $3 \pmod 8$.

(b) Prove that there are infinitely many primes of the form $8r + 3$, by taking k such primes, $p_1, \ldots, p_k$ and considering $Q = (p_1 \ldots p_k)^2 + 2$, arguing similarly to 2.2(b) above and using (a). (This is slightly simpler than 2.2(b) since Q is odd.)

(c) Suppose that $p = 4r+1$ and $q = 2p+1$ are both prime. Show that 2 is a primitive root mod q, that is $\mathrm{ord}_q\, 2 = 2p$. (As in 2.2(c) above, it is only necessary to eliminate $\mathrm{ord}_q\, 2 = p$; do this by showing it implies $\left(\tfrac{2}{p}\right) = 1$.)

(d) In Chapter 6,2.6(h), an indication is given of how to prove the following: if the (odd) prime p divides the Fermat number $F_k = 2^{2^k} + 1$, then $\mathrm{ord}_p 2 = 2^{k+1}$, which implies $2^{k+1} \mid (p-1)$. If $k \geq 2$, then the latter implies $p \equiv 1 \pmod 8$; now use 2.1 and Euler's criterion to show that in fact $2^{k+2} \mid (p-1)$. For example, any prime factor p of F_5 must satisfy $p \equiv 1 \pmod{128}$. In fact, as Euler discovered in 1732, though not using this result, $641 \mid F_5$. Unfortunately F_6 is just too big to handle safely with extended precision; otherwise you could look for the factor $274\,177 = 256 \cdot 1071 + 1$.

(e) Given that $p = 8n + 5$ is prime and $a^{(p-1)/4} \equiv -1 \pmod p$, show that $x = \pm 2^{2n+1} a^{n+1}$ satisfies $x^2 \equiv a \pmod p$. (Compare Chapter 8,2.4(g).) [Hint. Show that it is enough to prove $2^{4n+2} \equiv -1 \pmod p$. Use 2.1 above, and Euler's criterion 1.4(a).]

2.4 Computing exercise Find the primes $q < 2000$ for which 2.2(c) or 2.3(c) proves that 2 is a primitive root.

Proof of 2.1 (the quadratic character of 2)   Using Euler's criterion all we need is a systematic way of calculating $2^{(p-1)/2} \pmod p$ for odd primes p. Consider the numbers

$$2 \cdot 1, 2 \cdot 2, 2 \cdot 3, \ldots, 2 \cdot \frac{1}{2}(p-1) \qquad (*)$$

Claim   *These are all absolutely incongruent mod p, i.e. if $2r \equiv \pm 2s \pmod p$ where $1 \leq r, s \leq (p-1)/2$, then $r = s$.*

Proof   The hypotheses imply $r \equiv \pm s \pmod p$ and $-\frac{1}{2}(p-3) \leq r - s \leq \frac{1}{2}(p-3)$, so $r = s$.   □

Now every number t has a unique 'least absolute residue mod p', that is a number u congruent to t and satisfying $-\frac{1}{2}(p-1) \leq u \leq \frac{1}{2}(p-1)$. Replacing each number in $(*)$ by this least absolute residue mod p the resulting numbers form a new list $(**)$ in which all entries are *distinct in absolute value*, by the claim; hence their absolute values must be $1, 2, 3, \ldots, \frac{1}{2}(p-1)$ in some order. Multiplying all the numbers in $(*)$ together and cancelling $2, \ldots, \frac{1}{2}(p-1)$ now gives $2^{(p-1)/2} \equiv (-1)^m \pmod p$ where m is the number of minus signs in the list $(**)$. Thus $(-1)^m$ also equals the Legendre symbol required.   □

Example: $p = 7$ gives $2,4,6$ for $(*)$ and $2, -3, -1$ for $(**)$, so $m = 2$.

It is now a matter of looking at four cases according as $p = 8k + 1$, $p = 8k + 3$, $p = 8k + 5$ or $p = 8k + 7$ for some integer k. Let us look at one case and leave the others as exercises. If $p = 8k + 1$ then the list $(**)$ is

$$2 \cdot 1, 2 \cdot 2, 2 \cdot 3, \ldots, 2 \cdot 2k, -4k + 1, -4k + 3, \ldots, -1,$$

the first negative number here being $2(2k+1) - p$. Hence $m = 2k$, and $(-1)^m = 1$, as required.

Finally, the formula for $\left(\frac{2}{p}\right)$ follows since $(p^2-1)/8$ is even when $p \equiv \pm 1 \pmod 8$ and odd when $p \equiv \pm 3 \pmod 8$.

3. Quadratic reciprocity

This remarkable theorem, first proved by Gauss in 1796 and later re-proved by him and others dozens of times, relates the quantities $\left(\frac{p}{q}\right)$ and $\left(\frac{q}{p}\right)$ when p and q are distinct odd primes. It seems on the face of it that the answer to the question 'Is pRq?' is unlikely to influence the answer to the question 'Is qRp?' However this is far from being the case, as you might have guessed if you tried 1.7(c) above.

3.1 The law of quadratic reciprocity *If p and q are distinct odd primes, then $\left(\frac{p}{q}\right) = \left(\frac{q}{p}\right)$ unless both are congruent to 3 (mod 4). This is equivalent to the statement*

$$\left(\frac{p}{q}\right)\left(\frac{q}{p}\right) = (-1)^{(p-1)(q-1)/4}.$$

We leave you to check the equivalence of the two statements. The first observation is that the proof of 2.1 above goes over word for word to a proof of the following result:

3.2 Gauss's lemma *Let p be an odd prime and $p \nmid n$. Then $\left(\frac{n}{p}\right) = (-1)^m$, where m is the number of negative terms when the sequence*

$$n, 2n, 3n, \ldots, \frac{1}{2}(p-1)n \qquad\qquad (*)$$

is replaced by the sequence $(**)$ of least absolute residues of these numbers mod p (the residues between $-\frac{1}{2}(p-1)$ and $\frac{1}{2}(p-1)$ inclusive). Also, the absolute values of the numbers in $(**)$ are just $1, 2, \ldots, \frac{1}{2}(p-1)$ in some order.

Now let $s = \frac{1}{2}(p-1)$. We shall prove the following.

3.3 Lemma *Suppose n is odd, p is an odd prime, and s is as above. Let*

$$M = \left[\frac{n}{p}\right] + \left[\frac{2n}{p}\right] + \ldots + \left[\frac{sn}{p}\right].$$

Then $m \equiv M \pmod 2$, so that from 3.2, $\left(\frac{n}{p}\right) = (-1)^M$.

Proof of 3.3 Dividing p into jn for $1 \leq j \leq s$ gives

$$jn = p\left[\frac{jn}{p}\right] + r_j \text{ where } 0 < r_j < p.$$

Adding these j equations gives

$$\frac{p^2 - 1}{8} n = pM + r_1 + r_2 + \ldots + r_s.$$

Now m of these r_i are replaced by $r_i - p$ when passing from the sequence $(*)$ to the sequence $(**)$ of least absolute residues mod p, and then the absolute values of the numbers in $(**)$ become $1, 2, \ldots, \frac{1}{2}(p-1)$ in some order. If we work mod 2, then $-x \equiv x$ for any integer x, so we have

$$1 + 2 + \ldots + \frac{1}{2}(p - 1) \equiv r_1 + r_2 + \ldots + r_s + mp \pmod{2}.$$

The last two equations now give, on subtraction,

$$\frac{p^2 - 1}{8}(n - 1) \equiv (M - m)p \pmod{2}.$$

Now the left-hand side is even (this uses n *odd*), and p is odd, so that $M - m$ is even. $\qquad\square$

Remark When $n = 2$, what relationship does the above give between M and m? Compare the proof of 2.1 above.

The final part of the proof is an argument essentially due to Gauss's pupil F.G. Eisenstein (also famous for his criterion for irreducibility of a polynomial with integer coefficients). See also *College Math. J.* **25** (1994), 29–34.

Proof of 3.1 In the plane with coordinates x and y, mark the integers $1, 2, \ldots, \frac{1}{2}(p - 1)$ along the x-axis and the integers $1, 2, \ldots, \frac{1}{2}(q - 1)$ along the y-axis, as in Fig. 11.1, and form the rectangle with vertices $(0, 0)$, $(p/2, 0)$, $(p/2, q/2)$, $(0, q/2)$. The rectangle contains in its interior $\frac{1}{2}(p - 1)\frac{1}{2}(q - 1)$ *lattice points*, with both coordinates integers. Consider the diagonal of the rectangle; this has equation $y = xq/p$ with $0 \leq x \leq p/2$ and $0 \leq y \leq q/2$ and so cannot contain any of these lattice points.

The line $x = k$ ($1 \leq k \leq \frac{1}{2}(p - 1)$) contains the lattice points $(k, 1)$, $(k, 2), \ldots, (k, \left[\frac{kq}{p}\right])$ *below* the diagonal. Thus writing M for the total number of lattice points below the diagonal then by 3.3, with n replaced by q, we have $\left(\frac{q}{p}\right) = (-1)^M$.

Similarly if N is the number of lattice points *above* the diagonal then $\left(\frac{p}{q}\right) = (-1)^N$. Hence $\left(\frac{p}{q}\right)\left(\frac{q}{p}\right) = (-1)^{N+M}$ and since $N + M = \frac{1}{2}(p - 1)\frac{1}{2}(q - 1)$ this completes the proof of 3.1. $\qquad\square$

3.4 Examples

(a) We can illustrate both the strength and the weakness of 3.1 by taking the prime 1711 (compare 5.3 and 5.1 in Chapter 10). Thus let us work out $\left(\frac{1711}{p}\right)$ for some primes p.

$\left(\frac{1711}{3}\right) = \left(\frac{1}{3}\right)$ since $1711 \equiv 1 \pmod{3}$, and this is 1.

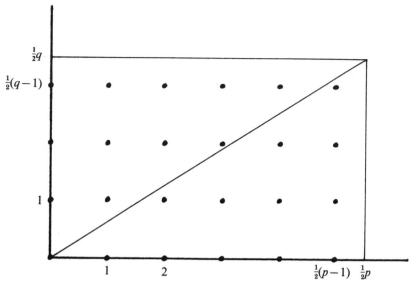

Fig. 11.1

Similarly
$$(\tfrac{1711}{5}) = (\tfrac{1}{5}) = 1.$$
Next,
$$(\tfrac{1711}{7}) = (\tfrac{3}{7}) = - (\tfrac{7}{3}) \text{ by reciprocity } = - (\tfrac{1}{3}) = -1.$$
$$(\tfrac{1711}{11}) = (\tfrac{6}{11}) = (\tfrac{2}{11})(\tfrac{3}{11}) \text{ (see 1.4(b))} = - (\tfrac{3}{11}) \text{ by 2.1}$$
$$= + (\tfrac{11}{3}) \text{ by reciprocity } = (\tfrac{2}{3}) = -1.$$

Note that 3 and 5 *do* occur among the prime factorizations of the Q^* in Chapter 10,5.3, whereas 7 and 11 *don't*.

So far, so good, but suppose we try a larger prime $p = 997$:
$$(\tfrac{1711}{997}) = (\tfrac{714}{997}) = (\tfrac{2}{997})(\tfrac{3}{997})(\tfrac{7}{997})(\tfrac{17}{997})$$
on factorizing $714 = 2 \cdot 3 \cdot 7 \cdot 17$. Now each of these can be evaluated, the first by 2.1 and the others using reciprocity and the other properties of 1.1 and 1.4 above. But this is clearly going to be a moderately lengthy procedure, and it has the great drawback of needing a factorization of 714 to begin with. Fortunately there is a much quicker way, using the amazing Jacobi symbol, which generalizes the Legendre symbol, and which we introduce in Section 4 below. So we postpone writing a program to evaluate the symbol $(\tfrac{n}{p})$ until Section 4. That should not prevent you, however, from using the results above to evaluate some Legendre symbols, as in the following exercises.

(b) It is relatively easy to work out all odd primes p for which, say, 3 is a quadratic residue. For $(\tfrac{3}{p}) = (\tfrac{p}{3}) = (\tfrac{1}{3}) = 1$ if $p \equiv 1 \pmod 4$ and

$p \equiv 1 \pmod 3$, i.e. if $p \equiv 1 \pmod{12}$. Similarly $\left(\frac{3}{p}\right) = \left(\frac{p}{3}\right) = \left(\frac{-1}{3}\right) = -1$ if $p \equiv 1 \pmod 4$ and $p \equiv -1 \pmod 3$, i.e. if $p \equiv 5 \pmod{12}$. The other two cases, with $p \equiv 3 \pmod 4$ work similarly and we deduce that

$$\left(\tfrac{3}{p}\right) = 1 \text{ if and only if } p \equiv \pm 1 \pmod{12}.$$

3.5 Exercises

(a) Suppose that $n \geq 1$ and $F = 2^{2^n} + 1$ is prime. Show that $F \equiv 5 \pmod{12}$, and deduce from 3.4(b) that $\left(\frac{3}{F}\right) = -1$. It follows that $k = 3$ is always a suitable number for use in Pepin's test for Fermat numbers (see 1.6(g) above). Show that 5 is also a suitable number, as is 10.

(b) Use Pepin's test (and the power algorithm program) to check the primality or otherwise of such Fermat numbers as are within extended precision (not many!).

(c) Show that $\left(\frac{5}{p}\right) = 1$, p being an odd prime, if and only if $p \equiv \pm 1 \pmod{10}$, and that $\left(\frac{7}{p}\right) = 1$ if and only if $p \equiv \pm 1$ or ± 3 or $\pm 9 \pmod{28}$. (For the latter, first check that the quadratic residues mod 7 are 1,2 and 4.) There are tables of results of this kind, for example Table 32 in [Riesel (1985)]. If you want some more practice, here are a few more results:

$$\left(\tfrac{-3}{p}\right) = 1 \text{ if and only if } p \equiv 1 \pmod 6,$$

$$\left(\tfrac{-5}{p}\right) = 1 \text{ if and only if } p \equiv 1, 3, 7 \text{ or } 9 \pmod{20},$$

$$\left(\tfrac{11}{p}\right) = 1 \text{ if and only if } p \equiv \pm 1, \pm 5, \pm 7, \pm 9 \text{ or } \pm 19 \pmod{44}.$$

(d) **Converse of Proth's theorem.** The following theorem includes most of Proth's theorem (Chapter 6,3.8) and also its converse:

Suppose that $n = k \cdot 2^m + 1$, $m \geq 2$, k odd and $k < 2^m$. Suppose that p is an odd prime with $\left(\frac{n}{p}\right) = -1$. Then

$$n \text{ is prime} \Leftrightarrow p^{(n-1)/2} \equiv -1 \pmod n.$$

Prove the $\Leftarrow$ part using Proth's theorem and the $\Rightarrow$ part using quadratic reciprocity and Euler's criterion.

Thus in Chapter 6,3.9, it is unnecessary to use Miller's test for checking the compositeness of certain of the numbers, so long as the choice of base is prime.

(e) Show in the following way that there are infinitely many primes of the form $5r + 4$. Let $Q = 5(n!)^2 - 1$, and suppose that p is a prime dividing Q. Show that $p \nmid n!$ and $\left(\frac{5}{p}\right) = 1$; deduce from (c) above that $p \equiv \pm 1 \pmod 5$. Now show that not *all* prime factors of Q can be $\equiv 1 \pmod 5$ and deduce the result.

(f) Several results above (such as (e)) are special cases of the famous theorem of G.L. Dirichlet (1837): If $(a, b) = 1$ then the arithmetic progression $ar + b$ ($r = 0, 1, 2, \ldots$) contains infinitely many primes.

(Are there any other special cases you could prove along the lines of (e)?). We cannot prove this theorem in full generality here, but here is a pleasant application (a problem from *Amer. Math. Monthly* **91**(1984), p. 521) which also uses quadratic reciprocity:

For any prime p, let g_p denote the *least positive primitive root* mod p. *Given any integer M, there exists infinitely many primes p for which $M < g_p < p - M$.*

(Thus there are infinitely many primes for which one needs a primitive root which is very large in absolute value.). Let $q_1 = 2$, $q_2 = 3$, $q_3 = 5, \ldots, q_n$ be the primes $\leq M$, and let $p = 1 + 8q_2q_3 \ldots q_n r$ be prime (by Dirichlet's theorem there are infinitely many such primes; in fact this special case, with '$a = 1$', is much easier to prove than the general theorem; see [Ribenboim (1988), p. 208.]) Show that $\left(\frac{-1}{p}\right) = 1$, $\left(\frac{2}{p}\right) = 1$ and, using quadratic reciprocity, that $\left(\frac{q_i}{p}\right) = 1$ for $i = 2, \ldots, n$. Deduce that all integers k with $-M \leq k + tp \leq M$, t being an arbitrary integer, are quadratic residues mod p, and so are not primitive roots mod p (they are *even* powers of primitive roots, by Chapter 8,2.5). Now deduce the result. (Equally, the greatest negative primitive root, G_p say, satisfies $-p + M < G_p < -M$.)

(g) Suppose that $p \equiv 1$ to the moduli 8 and all primes from 3 to 23 inclusive. Show that, for any prime $q \leq 23, \left(\frac{q}{p}\right) = 1$, and deduce that p has no quadratic nonresidue under 29 (compare 1.6(h) above). Note that in fact $1 + 8 \cdot 3 \cdot 5 \ldots 23 = 892\,371\,481$ *is* prime so this provides an example. Can you find any similar examples with other runs of consecutive primes?

We conclude this section with a proof of the part of the Lucas-Lehmer test for primality of Mersenne numbers $2^p - 1$; see 3.3 *et seq.* of Chapter 9 for the statement of the test and for the notation used here. The proof given here is essentially the same as that in [M.I. Rosen (1988)]. The object is to prove that, *if p is a prime ≥ 3 and M_p is prime, then M_p divides r_{p-1}.*

*3.6 Proof of the 'only if' part of the Lucas–Lehmer test, Chapter 9, 3.3

Let $\tau = (1 + \sqrt{3})/\sqrt{2}$ and $\overline{\tau} = (1 - \sqrt{3})/\sqrt{2}$ (so $\tau\overline{\tau} = -1$) and write $q = M_p$, assuming q is prime. Raising $\tau\sqrt{2} = 1 + \sqrt{3}$ to the qth power gives

$$\tau^q 2^{(q-1)/2}\sqrt{2} = 1 + q\sqrt{3} + \binom{q}{2}3 + \binom{q}{3}3\sqrt{3} + \ldots + (\sqrt{3})^q$$

$$= 1 + aq\sqrt{3} + bq + 3^{(q-1)/2}\sqrt{3},$$

for integers a and b, using the fact that the binomial coefficients $\binom{q}{k}$, for $1 \leq k \leq q-1$, are all divisible by the prime q (see Chapter 1, 2.3(d)).

Since $q = M_p = 2^p - 1$, and $p \geq 3$, we have $q \equiv -1 \pmod 8$. Hence, by 2.1 above, 2 is a quadratic residue mod q. Hence by Euler's criterion 1.3, $2^{(q-1)/2} \equiv 1 \pmod q$. Also, $2^p \equiv -1 \pmod 3$ since p is odd, so $q \equiv 1 \pmod 3$. Using quadratic reciprocity (3.1 above) we have

$$\left(\tfrac{3}{q}\right) = -\left(\tfrac{q}{3}\right) = -\left(\tfrac{1}{3}\right) = -1,$$

so that by Euler's criterion $3^{(q-1)/2} \equiv -1 \pmod q$ Substituting in the above formula then gives

$$\tau^q(1 + cq)\sqrt{2} = 1 + aq\sqrt{3} + bq + (-1 + dq)\sqrt{3},$$

where c and d are integers. Hence

$$\tau^q(1 + cq) = \overline{\tau} + q(a\sqrt{3} + b + d\sqrt{3})/\sqrt{2},$$

and multiplying through by τ we deduce

$$\tau^{q+1}(1 + cq) = -1 + q(e + f\sqrt{3})/2,$$

for some integers e and f.

Now $\tau^2 = \omega$ (see the proof of the other half of Chapter 9, 3.3) and $\tau^{q+1} = \tau^{2^p}$ so the last equation gives

$$\omega^{2^{p-1}}(1 + cq) = -1 + q(e + f\sqrt{3})/2.$$

Multiplying both sides by $2\overline{\omega}^{2^{p-2}}$ and using the fact that $\omega\overline{\omega} = 1$ we deduce after some rearrangement, and using 3.5 of Chapter 9, that

$$2r_{p-1} = \overline{\omega}^{2^{p-2}} q(e + f\sqrt{3}) - 2\omega^{2^{p-2}} cq.$$

If the right-hand side is multiplied out (in principle if not in practice!) then the $\sqrt{3}$ terms must cancel, since $\sqrt{3}$ is irrational, and the equation tells us, since q is odd, that $q \mid r_{p-1}$, as required. $\square$

4. The Jacobi symbol and a program for finding $\left(\frac{n}{p}\right)$

In this section we introduce a generalization of the Legendre symbol which makes the calculation of $\left(\frac{n}{p}\right)$ much easier since no factorization of n is necessary.

4.1 Definition of the Jacobi symbol   *Let k be odd and > 0, let $(n, k) = 1$, and suppose that k is a product of (not necessarily distinct) primes: $k = p_1 p_2 \ldots p_r$. Then*

$$\left(\tfrac{n}{k}\right) = \left(\tfrac{n}{p_1}\right)\left(\tfrac{n}{p_2}\right) \ldots \left(\tfrac{n}{p_r}\right).$$

Note that all these Legendre symbols are defined, since no p_i can divide n (k and n are coprime). Thus $\left(\frac{n}{k}\right) = \pm 1$ and, if k happens to be

prime, then the Jacobi and Legendre symbols coincide. Note also that
if nRk, then certainly $\left(\frac{n}{p_i}\right) = 1$ for each i, and consequently $\left(\frac{n}{k}\right) = 1$.
However, the converse is not true as you can verify by showing that

$$\left(\tfrac{2}{15}\right) = \left(\tfrac{2}{3}\right)\left(\tfrac{2}{5}\right) = (-1)(-1) = 1,$$

while $x^2 \equiv 2 \pmod{15}$ has no solutions.

This makes the Jacobi symbol look less than useful, but the good
news is that the Jacobi symbol shares several crucial properties with the
Legendre symbol, and these are enough to make it very useful indeed.

4.2 Proposition: basic properties of the Jacobi symbol (compare 1.4)

(a) *If k is odd and > 0 and $(k, n) = (k, m) = 1$, then $\left(\frac{n}{k}\right)\left(\frac{m}{k}\right) = \left(\frac{nm}{k}\right)$.*

(b) *If k is odd and > 0 and $(k, n) = 1$, then $\left(\frac{n}{k}\right) = \left(\frac{n+ik}{k}\right)$ for any integer i.*

Proof These follow immediately from the definition of the Jacobi symbol together with the corresponding properties of the Legendre symbol.

□

4.3 Proposition (compare 2.1 above) *Let k be odd and > 0. Then $\left(\frac{2}{k}\right) = 1$ if and only if $k \equiv \pm 1 \pmod 8$. (The remaining odd numbers, for which $\left(\frac{2}{k}\right) = -1$, are of course $k \equiv \pm 3 \pmod 8$.) This can be equivalently stated as*

$$\left(\tfrac{2}{k}\right) = (-1)^{(k^2-1)/8}$$

Proof This is a matter of carefully using the result for k prime (2.1 above). Write $k = p_1 p_2 \dots p_s p_{s+1} \dots p_r$, a product of not necessarily distinct primes, all odd, and with the primes $\equiv \pm 3 \pmod 8$ being $p_1, \dots, p_s$. It is easy to check that $k \equiv \pm 1 \pmod 8$ if and only if s is even. Now,

$$\left(\tfrac{2}{k}\right) = \left(\tfrac{2}{p_1}\right)\left(\tfrac{2}{p_2}\right)\dots\left(\tfrac{2}{p_r}\right) = 1 \Leftrightarrow \text{ an even number of the } \left(\tfrac{2}{p_i}\right) \text{ are } -1$$

$$\Leftrightarrow \text{ an even number of } p_i \text{ are } \equiv \pm 3$$

$$\pmod 8 \text{ (using 2.1)}$$

$$\Leftrightarrow k \equiv \pm 1 \pmod 8, \text{ by the above.} \qquad \square$$

4.4 Theorem: reciprocity for the Jacobi symbol (compare 3.1)
If n and k are odd, positive and coprime then $\left(\frac{n}{k}\right) = \left(\frac{k}{n}\right)$ unless n and k are both congruent to 3 (mod 4) This is equivalent to the statement

$$\left(\tfrac{n}{k}\right)\left(\tfrac{k}{n}\right) = (-1)^{(n-1)(k-1)/4}.$$

Proof Again, this is just a matter of carefully using the corresponding result 3.1 which holds when n and k are distinct primes. In fact, we factorize n and k into odd primes, say

$$k = p_1 p_2 \ldots p_s p_{s+1} \ldots p_r, n = q_1 q_2 \ldots q_u q_{u+1} \ldots q_v,$$

where (for later use) $p_1, \ldots, p_s$ and $q_1, \ldots, q_u$ are $\equiv 3 \pmod 4$ and the others are $\equiv 1 \pmod 4$. Then application of the definition of the Jacobi symbol and of the basic property 4.2(a), together with the formula in 3.1 for $\left(\frac{p}{q}\right)\left(\frac{q}{p}\right)$ when p and q are distinct primes, shows that

$$\left(\tfrac{n}{k}\right)\left(\tfrac{k}{n}\right) = (-1)^N, \text{ where } N = \sum_{i,j} \frac{p_i - 1}{2}\frac{q_j - 1}{2}.$$

Here the sum is over $i = 1, \ldots, r$ and $j = 1, \ldots, v$.

We have to prove that N is even unless n and k are both $\equiv 3 \pmod 4$, that is unless s and u are both odd. Noting that $\frac{p-1}{2}$ is even if and only if $p \equiv 1 \pmod 4$, we have

N is even $\Leftrightarrow$ there are an even number of odd summands

       $\Leftrightarrow$ there are an even number of pairs (i, j)

           with p_i and q_j *both* $\equiv 3 \pmod 4$

       $\Leftrightarrow$ su is even

       $\Leftrightarrow$ s is even or u is even.   □

To show how this enables us to find Legendre symbols far more quickly than before, consider again the example $\left(\frac{1711}{997}\right)$ considered in 3.4(a) above. We have

$$\left(\tfrac{1711}{997}\right) = \left(\tfrac{714}{997}\right) \text{ by 4.2(b)}$$
$$= \left(\tfrac{2}{997}\right)\left(\tfrac{357}{997}\right) \text{ by 4.2(a)}$$
$$= -\left(\tfrac{357}{997}\right) \text{ by 4.3}$$
$$= -\left(\tfrac{997}{357}\right) \text{ by 4.4}$$
$$= -\left(\tfrac{283}{357}\right) \text{ by 4.2(b)}$$
$$= -\left(\tfrac{357}{283}\right) \text{ by 4.4}$$
$$= -\left(\tfrac{74}{283}\right) \text{ by 4.2(b)}$$
$$= -\left(\tfrac{2}{283}\right)\left(\tfrac{37}{283}\right) \text{ by 4.2(a)}$$
$$= \left(\tfrac{37}{283}\right) \text{ by 4.3}$$
$$= \left(\tfrac{283}{37}\right) \text{ by 4.4}$$
$$= \left(\tfrac{24}{37}\right) \text{ by 4.2(b)}$$
$$= \left(\tfrac{2}{37}\right)^3\left(\tfrac{3}{37}\right) \text{ by 4.2(b)}$$
$$= -\left(\tfrac{3}{37}\right) \text{ by 4.3}$$
$$= -\left(\tfrac{37}{3}\right) \text{ by 4.4}$$
$$= -\left(\tfrac{1}{3}\right) \text{ by 4.2(b)}$$
$$= -1.$$

Hence 1711 is a quadratic nonresidue mod 997. If we apply the same procedure to $(\frac{n}{k})$ where k is *not prime*, and find the answer -1, then we *can* deduce that nNk, but if we find the answer 1 we *cannot* deduce nRk; compare the remarks following 4.1 above.

4.5 Exercises

(a) Work out the following by hand: $(\frac{1711}{1999})$, $(\frac{1711}{2007})$, $(\frac{1711}{2021})$, $(\frac{1711}{2027})$. For which denominators k can you deduce for sure whether $1711Rk$?

(b) In the course of the calculation of $(\frac{n}{k})$ the 'numerator' decreases. Why does it always eventually reach 1 or a power of 2?

(c) Show that (for k odd) $(\frac{-1}{k}) = 1$ if and only if $k \equiv 1 \pmod 4$. (A similar method to that of 4.3 works well.)

(d) Suppose that j and k are odd and > 0, that $j \equiv k \pmod n$, $(j, n) = 1$, and $n \equiv 0$ or $1 \pmod 4$. Show that $(\frac{n}{j}) = (\frac{n}{k})$. (Use the reciprocity theorem 4.4 on each side of this equation to begin with. When $n \equiv 0 \pmod 4$ you will find it helpful to use the formula for $(\frac{2}{k})$ given in 4.3 above.)

(e) Suppose that j and k are odd and > 0, that $j \equiv k \pmod{4n}$ and $(j, n) = 1$. Deduce from (d) above that $(\frac{n}{j}) = (\frac{n}{k})$. (Replace n by $4n$ in (d).) This useful result shows why, in 3.5(c) above, the value of $(\frac{n}{p})$ for an odd prime p depends only on the residue of p modulo $4n$. For $n = 7$, -5 and 11 the modulus $4n$ cannot be reduced, whereas for $n = 5$ and -3 it can be reduced to $2n$.

It is relatively easy to write a program to implement the method of evaluating $(\frac{n}{k})$ used in the above example. We continually reduce n mod k, take out any powers of 2 from n, reverse n and k and repeat. You may like to write your own program, but we give one possible implementation below. It does not include any safeguard against the possibility that n and k have a common factor (what will happen to the above algorithm then?), or that k is even. We also assume that $n > 0$.

The program prints out the successive Jacobi symbols occurring in the calculation, with a $+$ or $-$ sign in front as appropriate. Each line written in the program equals the original $(\frac{n}{k})$.

4.6 Program for evaluating the Jacobi symbol $(\frac{n}{k})$ for n and $k > 0$

```
PROGRAM P11_4_6;
VAR
count:  integer;
x, y, n, nsaved, k, ksaved, tempk :  extended;
sign:  boolean;
(* Sign records whether there is a + or - in front of *)
```

```
(* the Jacobi symbol at any stage of the calculation *)

  PROCEDURE WriteRatio (n, k:  extended;
       sign :  boolean);
  (* Displays the intermediate values of (n/k) *)
  (* with appropriate sign.  *)
  BEGIN
    writeln;
    IF NOT sign THEN
      write (' - (', n:0:0, '/', k:0:0,')')
    ELSE
      write (' (', n:0:0, '/', k:0:0,')');
  END (* of WriteRatio *);

BEGIN (* main program *)
  writeln ('Type the value of n');
  readln (n);
  nsaved:=n;
  writeln ('Now type the first value of k');
  readln (k);
  ksaved:=k;
  WHILE ksaved>2 DO
  (* Inputting k ≤ 2 later will end this loop *)
    BEGIN
      sign:=TRUE;
      WHILE n>1 DO
        BEGIN
          n:=n - k*INT(n/k);
          (* This is where we use 4.2(b) *)
          WriteRatio (n,k,sign);
          count:=0;

          WHILE n/2=INT(n/2) DO
            BEGIN
              n:=n/2;
              count:=count + 1;
            END;
          IF count/2 <> INT(count/2) THEN
            BEGIN
              x:=k - 8*INT(k/8);
              IF (x=3) OR (x=5) THEN
                sign:= NOT sign;
```

```
                    (* This is where we use 4.3 *)
            END;
         IF count>0 THEN
           WriteRatio (n,k,sign);
         IF n>1 THEN
           BEGIN
             xtempk:=k;
             k:=n;
             n:=tempk;
             (* Now n and k are interchanged *)
             x:=k - 4*INT(k/4);
             y:=n - 4*INT(n/4);
             IF (x=3) AND (y=3) THEN
                sign:= NOT sign;
                (* This is where we use 4.4 *)
             WriteRatio (n,k,sign);
           END;
       END; (* of WHILE n>1*)
       writeln;
       IF (sign = TRUE) THEN
         writeln('Hence (',nsaved:0:0,'/',ksaved:0:0,')
         = 1')
       ELSE
         writeln('Hence (',nsaved:0:0,'/',ksaved:0:0,')
         = - 1');
       writeln ('Type the next value of k
       (a value less than 3 ends the run)');
       readln (k);
       ksaved:=k;
       n:=nsaved
     END; (* of WHILE ksaved>2 loop *)
  END. (* of main program *)
```

4.7 Computing exercises   (a) Taking for n the numbers in 5.6 of Chapter 10, check that the primes p occurring in the factorization of the Q^*s are indeed those for which $\left(\frac{n}{p}\right) = 1$ (see 5.1 of Chapter 10).

(b) Check the examples of 4.5 above using the program P11_4_6.

(c) Amend the program so that it can handle $n < 0$. (One way is to write $\left(\frac{n}{k}\right) = \left(\frac{-1}{k}\right)\left(\frac{-n}{k}\right)$ and to use 4.5(c) above, but you can of course be more subtle if you wish. Beware that the INT function in PASCAL tends to round towards zero.)

(d) Redo the exercises 1.7(a) and (b) above, using the program.

4.8 Project   By combining programs P10_5_5 and P11_4_6, write a version of the method of Chapter 10 for factorizing a number n which, when seeking to factorize the Q^*s, uses only those primes $p <$ maxprime for which $\left(\frac{n}{p}\right) = 1$. Presumably once n is declared, the program will immediately work out those primes p for which $\left(\frac{n}{p}\right) = 1$ and for which p is less than some reasonably large maximum. These primes will then replace the file 'primes.dat'.

5. Solving the equation $x^2 \equiv a$ (mod p): quadratic congruences

The method presented in Section 4 is a very efficient one for determining whether the equation $x^2 \equiv a$ (mod p), where p is an odd prime, has a solution for x. (Of course if $p = 2$, or if $p \mid a$, then there is always a solution for x.) We turn here to the obvious and very basic question of finding an explicit x when a and p are given, and it is known that $\left(\frac{a}{p}\right) = 1$. We have already touched on this in Chapter 8,2.4(f), (g), where p is assumed to have the form $4n+3$ or $8n+5$. In the latter case we filled in a case not covered in Chapter 8, in 2.3(e) above.

There are a number of ways of finding an explicit solution to $x^2 \equiv a$ (mod p) for arbitrary a and p with $\left(\frac{a}{p}\right) = 1$; we give here (with some details left as exercises) an iterative method due to [Shanks (1972)] which is particularly suited to numerical computation. Despite its apparent complexity, it is extremely simple to program: see P11_5_5 below. Before giving the method we show that solving a *general quadratic congruence* $ax^2 + bx + c \equiv 0$ (mod p), where p is an odd prime, is no harder. We proceed very much as in the normal method of solving a quadratic equation by completing the square. Clearly we can assume $p \nmid a$ or we are back to linear congruences. Thus let $\overline{2}, \overline{a}$ be the inverses of $2, a$, respectively. Then the given congruence is equivalent to

$$(x + \overline{2a}b)^2 \equiv (\overline{2a})^2(b^2 - 4ac) \text{ (mod } p).$$

The number $b^2 - 4ac$ is called the *discriminant* of the quadratic congruence. Assuming that p does not divide the discriminant, the congruence has a solution if and only if $\left(\frac{b^2-4ac}{p}\right) = 1$, and we can find the solution by solving $y^2 \equiv b^2 - 4ac$ (mod p) for y and using $x \equiv \overline{2a}(-b\pm y)$, exactly as in the usual formula for solving a quadratic equation. Note that these solutions are *necessarily distinct* (mod p).

If $p \mid (b^2 - 4ac)$, but $p \nmid a$, then there is *one* solution $x \equiv -\overline{2a}b$ (mod p).

5.1 Exercises (a) Use program P11_4_6 to decide which of the following quadratic congruences have solutions.

$$100x^2 + 1000x + 1 \equiv 0 \pmod{p}$$

where $p = 1987, 1993, 1997, 1999, \ldots$ (consecutive primes, which you can find using program P2_2_1). How far does the 'pattern' of existence/non-existence persist? Which additional primes p, dividing the discriminant but not dividing $a = 100$, guarantee a single solution?

(b) **Fibonacci sequences mod p.** We use the notation of Chapter 3, 1.10, except that m will now be p (an odd prime). Show that $x^2 - x - 1 \equiv 0 \pmod{p}$ has a solution if and only if $p = 5$ or $p \equiv \pm 1 \pmod{10}$. (Use the first result of 3.5(c) above.)

We consider $p \equiv \pm 1 \pmod{10}$. Note that the solutions r, s are then *distinct* mod p, from the above remarks. Show that $r + s \equiv 1, rs \equiv -1$ and $(r-s)^2 \equiv 5 \pmod{p}$. Show by induction on n that $(r-s)f_n \equiv r^n - s^n \pmod{p}$, which determines $f_n \pmod{p}$ since $r \not\equiv s \pmod{p}$.

Suppose that both $r^j \equiv 1$ and $s^j \equiv 1 \pmod{p}$, where $j > 0$. Show that $f_j \equiv 0$ and $f_{j+1} \equiv 1 \pmod{p}$, so that the Fibonacci sequence mod p repeats from f_j onwards. Deduce that the period $k(p)$ (see Chapter 3, 1.10) satisfies $k(p) \mid j$. Taking $j = p - 1$ (a valid choice by Fermat's theorem) we deduce $k(p) \mid (p - 1)$.

It can also be shown that, when $p \equiv \pm 3 \pmod{10}$ (so $\left(\frac{5}{p}\right) = -1$), we have $k(p) \mid (2p + 2)$, but now the congruence $x^2 - x - 1 \equiv 0 \pmod{p}$ has no solution, so there is (presumably!) no such convenient formula for f_n. See [Wall (1960).]

Let now p be a prime with $p - 1 = 2^{s_0}(2k+1)$. Given a with $\left(\frac{a}{p}\right) = 1$ we seek to find x with $x^2 \equiv a \pmod{p}$. The method assumes that we also know an integer z with $\left(\frac{z}{p}\right) = -1$; this is easy to find by trial, starting with $z = 2$ and using the program P11_4_6 above. (See 1.6(b) above for a nasty example of this!) In what follows, all congruences are mod p unless otherwise stated.

We start with $x_0 \equiv a^{k+1}$, $n_0 \equiv a^{2k+1}$, which will be calculated using the power rule as in P4_2_3. Note that $x_0^2 \equiv a n_0$ so that if $n_0 \equiv 1$ then we are finished! Note that this is exactly what happens when $p \equiv 3 \pmod 4$ (compare Chapter 8, 2.4(f), and also (g)). In fact Shanks claims that it happens in about two-thirds of all cases; perhaps you can test this claim for yourself.

Otherwise, we shall find numbers b_0, b_1, $b_2, \ldots$ and define, for $i = 0, 1, 2, \ldots$:

$$x_{i+1} \equiv x_i b_i, \quad n_{i+1} \equiv n_i b_i^2.$$

It is easy to check by induction that $x_i^2 \equiv a n_i$ for all i. The numbers

b_i will be so chosen that the orders $\pmod{p}$ of $n_0, n_1, n_2, \ldots$ decrease strictly, therefore eventually becoming 1. When this happens we have the required solution x_i since $x_i^2 \equiv a n_i \equiv a$. In fact all the orders will be powers of 2: $\operatorname{ord} n_i = 2^{s_{i+1}}$ where $s_{i+1} = s_i - t_i$ and $t_i > 0$. Here and subsequently orders are mod p .

Let g be a primitive root mod p, so that $\operatorname{ord} g = p - 1$. Let $h = g^{2k+1}$; then $\operatorname{ord} h = 2^{s_0}$ and we consider the set $H \pmod{p}$ generated by h:

$$H = \{1, h, h^2, \ldots, h^{2^{s_0}-1}\}.$$

(For those familiar with the idea of a group, this is a subgroup of the group generated by g, the latter consisting of all nonzero residue classes mod p.) Note that as $h = g^{2k+1}$, the power $h^i = (g^{2k+1})^i = (g^i)^{2k+1}$ and g^i runs through all nonzero residue classes mod p, so H consists precisely of all nonzero $(2k+1)$th powers $\pmod{p}$. In particular $n_0 \in H$. Also, any element $m \in H$ satisfies $\operatorname{ord} m \mid 2^{s_0}$, so *the order of any element of H is a power of* 2.

We consider $\operatorname{ord} n_0$. Certainly as $n_0 \in H$, $\operatorname{ord} n_0$ is a power of 2, say $2^{s_0 - t_0}$ where $t_0 \geq 0$. We want to show that $t_0 > 0$. It is enough to show that n_0 fails to generate H, i.e., that some elements of H fail to be powers of n_0. In fact powers of n_0 must necessarily be *even* powers of h, for the following reason. Since a is a quadratic residue mod p, it is an *even* power of the primitive root g (Chapter 8,2.5 again); hence n_0, being a power of a, is also an even power of g, and powers of n_0 will all give even powers of g. These will therefore be even powers of h.

Hence:

5.2 Lemma $\operatorname{ord} n_0 = 2^{s_0 - t_0}$, *where $t_0 > 0$. We write $s_1 = s_0 - t_0$.*
   Recall that z is a quadratic nonresidue $\pmod{p}$.

5.3 Lemma $c_0 \equiv z^{2k+1}$ *generates H, that is, H consists precisely of the powers of c_0. Hence n_0 is a power of c_0: say $n_0 = c_0^{u_0}$.*

Proof   It is enough to show that $c_0 \in H$ and $\operatorname{ord} c_0 = 2^{s_0}$, for then the elements c_0^i for $i = 0, 1, \ldots, 2^{s_0}$ are in H and are distinct; hence they are precisely the elements of H. The fact that $c_0 \in H$ follows from the above remark that H consists of nonzero $(2k+1)$th powers.

Next, note that z is an *odd* power of g, since it is a quadratic nonresidue (compare 2.5 of Chapter 8). Hence, c_0 is an *odd* power of h, $c_0 \equiv h^{2j+1}$ say. Since $\operatorname{ord} c_0 \mid \operatorname{ord} h = 2^{s_0}$, the order of c_0 is certainly a power of 2, say 2^w where $w \leq s_0$. If $w < s_0$ then $(h^{2j+1})^{2^w} \equiv 1$ gives $2^{s_0} \mid (2j+1)2^w$, which is impossible. Hence the order of c_0 is 2^{s_0}, as required.   □

If $s_0 - t_0 = 0$ then we have $n_0 \equiv 1$ which, as pointed out above, finishes the search for x with $x^2 \equiv a$. So assume that $s_0 > t_0$ and define

$$b_0 \equiv c_0^{2^{t_0}-1} \in H, \; x_1 \equiv x_0 b_0, \; n_1 \equiv n_0 b_0^2 \in H, \; c_1 \equiv b_0^2 \in H.$$

Using $\operatorname{ord} c_0 = 2^{s_0}$ (see 5.3), it follows from these definitions that c_1 has order $2^{s_1} (s_1 = s_0 - t_0$ as in 5.2).

Just as n_0 is a power of c_0, we find n_1 is a power of c_1. We can also show that $\operatorname{ord} n_1 = 2^{s_2}$, say (necessarily a power of 2 since $n_1 \in H$), is *less than* $\operatorname{ord} n_0 = 2^{s_1} (= \operatorname{ord} c_0)$, that is $s_2 = s_1 - t_1$ where $t_1 > 0$. The argument for these two is contained in the following exercises (5.4); the reader may wish to assume these for the moment in order to see how the algorithm continues.

5.4 Exercises

(a) Using the fact that c_0 has order 2^{s_0}, and $n_0 \equiv c_0^{u_0}$ (see 5.3) has order $2^{s_0-t_0}$, show that 2^{t_0} is the exact power of 2 dividing u_0: $u_0 = 2^{t_0}(2m_0 - 1)$ say.

(b) Show that $n_1 \equiv c_1^{2m_0}$: a power of c_1.

(c) Show that the 2^{s_1-1} power of n_1 is $\equiv 1$ by expressing it as a power of c_0. Deduce that the order of n_1 is $< 2^{s_1}$. (If m_0 is odd you can deduce that it is exactly 2^{s_1-1}.)

As noted above, we have $x_1^2 \equiv a n_1$ so if $n_1 \equiv 1$ we are finished. If $n_1 \not\equiv 1$, that is if $s_1 > t_1$, then we iterate the process:

$$b_1 \equiv c_1^{2^{t_1}-1} \in H, \; x_2 \equiv x_1 b_1, \; n_2 \equiv n_1 b_1^2 \in H, \; c_2 \equiv b_1^2 \in H.$$

Since n_2 and c_2 are in H, their orders are powers of 2; we find $\operatorname{ord} c_2 = 2^{s_2}$ from the definition, using $\operatorname{ord} c_1 = 2^{s_1}$. The same argument as before will now show that $\operatorname{ord} n_2 < \operatorname{ord} n_1 (= \operatorname{ord} c_2)$, $\operatorname{ord} n_2 = 2^{s_3}$ say, where $s_3 = s_2 - t_2$ and $t_2 > 0$, and also that n_2 is a power of c_2. If $n_2 \equiv 1$ then we are finished; otherwise we iterate again. The order of n_i, being strictly decreasing, must reach 1, at which point $n_i = 1$ and $x = x_i$ is the required solution.

To implement this in a program is surprisingly easy, as Shanks points out in his article [Shanks (1972)]. We assume given the numbers s_0 and k (coming from $p-1 = 2^{s_0}(2k+1)$), together with a (such that aRp) and z (such that zNp). We determine $x_0 \equiv a^{k+1}$, $n_0 \equiv a^{2k+1}$ and $c_0 \equiv z^{2k+1}$ using the power algorithm.

Then n_0 is squared until it becomes $\equiv 1$ (this needs $s_1 = s_0 - t_0$ squarings). The value of s_1 is recorded.

We then take c_0 and square it $t_0 - 1$ times to get b_0.

By a happy chance, if the loop carrying out these squarings includes one pass through to change the object being squared (y in the program below) from n_0 to c_0, then the total number of passes through the loop is $s_0 - t_0 + 1 + t_0 - 1 = s_0$. This greatly simplifies the program.

Finally we evaluate $c_1 \equiv b_0^2$, $x_1 \equiv x_0 b_0$ and $n_1 \equiv n_0 c_0$, and continue (if $n_1 \neq 1$), increasing all subscripts in the above by 1.

You may like to write your own program to implement the above algorithm; here is our version.

5.5 Program for solving $x^2 \equiv a$ (mod p)

```
PROGRAM P11_5_5;
VAR
a,b,c,k1,odd,n,p,p1,x,y,z :  extended;
i,index,looplength,s:  integer;

   PROCEDURE PowerRule(base,power,modulus :  extended;
   VAR res :  extended);
```

This is exactly as in P4_2_3

```
BEGIN    (* of main program *)
  writeln('type p,a and z');
  readln(p,a,z);
  p1:=p- 1;
  index:=0;
  WHILE (p1/2 = INT(p1/2)) DO
    BEGIN
       p1:=p1/2;
       index:=index+1;
    END;
  s:=index;
  odd:=p1;  (* odd is the number 2k+1 in the text *)
  PowerRule(z,odd,p,c);    (* working out c *)
  PowerRule(a,odd,p,n);    (* working out n *)
  writeln('n:=',n:0:0);
  (* Not essential, but nice to see the n's *)
  k1:=(odd+1)/2;  (* this is k+1 *)
  PowerRule(a,k1,p,x);    (* working out x *)
  writeln('x:=',x:0:0);    (* This will be the first x *)
  WHILE (n>1) DO
    BEGIN
       looplength:=s;
       y:=n;   (* y will be the variable that is *)
       (* squared *)
       FOR i:=1 TO looplength DO
```

```
BEGIN
  IF (y=1) THEN
    BEGIN
      y:=c;
      s:=i- 1;
    END
  ELSE
    BEGIN
      y:=y*y;
      y:=y- p*INT(y/p);
    END;
  END;
  b:=y;
  c:=b*b; c:=c- p*INT(c/p);
  x:=b*x; x:=x- p*INT(x/p);
  writeln('x=',x:0:0);    (* Keeping track of x *)
  n:=c*n; n:=n- p*INT(n/p);
  writeln('n=',n:0:0);
  (* Again nice to keep track of n *)
  END;
writeln('x =',x:0:0);
(* This will be the final value of x *)
readln;
END.
```

5.6 Computing exercises

(a) Find a small quadratic nonresidue mod p for the primes 17, 19, 997, 1 000 333, 10 000 993, using P11_4_6. Hence solve the congruences

$$x^2 \equiv 15 \ (\text{mod } 17), \ x^2 \equiv 11 \ (\text{mod } 19), \ x^2 \equiv 589 \ (\text{mod } 997) \ ,$$

$$x^2 \equiv 3 \ (\text{mod } 1\,000\,333), \ x^2 \equiv 2 \ (\text{mod } 10\,000\,993).$$

(b) Investigate Shanks's claim that on average two-thirds of cases require no iteration at all: x_0 is the solution. (Recall from Chapter 8, 2.4(f) that when $p \equiv 3 \ (\text{mod } 4)$ this *always* happens.) See [Turner (1994)].

(c) Why is the maximum possible number of steps in Shanks's algorithm equal to s_0? Here are some examples where a (rather easy!) congruence requires the maximum number of iterations; perhaps you can find more. In each case the nonresidue chosen is 3, and the congruence is $x^2 \equiv 9$.

$$\text{mod } 17: \ s_0 = 4; \ \text{mod } 257: \ s_0 = 8;$$

$$\text{mod } 65\,537: s_0 = 16; \ \text{mod } 167\,772\,161: \ s_0 = 25.$$

5.7 Exercises: solution of $x^2 \equiv a$ (mod p^k), p an odd prime

(a) Suppose that p is an odd prime, $k \geq 2, p \nmid a$ and x_1 is a solution of
$$x^2 \equiv a \text{ (mod } p^k), \qquad 0 < x < p^k. \tag{1}$$
Show that x_1 can be written uniquely as $x_1 = x_2 + sp^{k-1}$ where $0 \leq s < p$ and x_2 is a solution of
$$x^2 \equiv a \text{(mod } p^{k-1}), \qquad 0 < x < p^{k-1}. \tag{2}$$

(b) Suppose that x_2 satisfies (2). Show that $x_1 = x_2 + sp^{k-1}$ satisfies (1) if and only if
$$2x_2 s \equiv -\left(\frac{x_2^2 - a}{p^{k-1}} \right) \text{(mod } p).$$
Here the right-hand side (*not* a Jacobi symbol!) is an integer because x_2 satisfies (2). Deduce that x_1 is uniquely determined by x_2.

(c) Deduce that (1) has the same number of solutions as $x^2 \equiv a$ (mod p), namely two if $\left(\frac{a}{p} \right) = 1$ and none otherwise.

(d) What happens if $p \mid a$?

5.8 Computing exercise Write a program, based on P11_5_5, for solving $x^2 \equiv a$ (mod p^2) when $\left(\frac{a}{p} \right) = 1$. You will find the program of Chapter 1,3.17(a) (or 3.16) helpful in calculating the inverse of $2x_2$ (mod p) in the above calculation.) You could be more ambitious and solve $x^2 \equiv a$ (mod p^k), of course!

Solve the congruences where p is replaced by p^2 in 5.6(a) above. The last of these,
$$x^2 \equiv 2 \text{ (mod } 10\,000\,993^2 = 100\,019\,860\,986\,049)$$
is accessible to extended precision, though you will need to use Head's algorithm (Chapter 4,3.4) to check that the answer is correct! The answer comes to $x = 33\,406\,487\,105\,406$ (and $s = 3\,340\,317$).

5.9 Exercises: solution of $x^2 \equiv a$ (mod 2^k) The method given in 5.7 does not work here, but a simple modification does the trick. When $k = 1$ or 2 we leave the solution to you! Also it is enough to consider the case when a is odd (otherwise k can be reduced), which implies that all solutions are odd also. From now on *we assume that a is odd and $k \geq 3$.*

(a) Suppose that
$$x^2 \equiv a \text{ (mod } 2^k). \tag{3}$$
Using the fact that any square is congruent to 1 (mod 8), show that $a \equiv 1$ (mod 8). *From now on we assume that this holds.*

(b) Show that if x and y are both solutions of (3), then $\frac{x-y}{2} \cdot \frac{x+y}{2} \equiv 0$ (mod 2^{k-2}). Deduce that 2^{k-2} divides one of the factors and hence that

$x \equiv \pm y \pmod{2^{k-1}}$. Deduce that if there is one solution x of (3) then there are exactly four solutions $\pmod{2^k}$, namely $\pm x$ and $\pm x + 2^{k-1}$.

(c) Starting with the solution $x = 1$ of (3) when $k = 3$ (remember $a \equiv 1 \pmod 8$), construct a solution of (3) with any k. [Hint. Let x_2 satisfy $x^2 \equiv a \pmod{2^{k-1}}$. Instead of writing, as in 5.7(a), $x_1 = x_2 + s \cdot 2^{k-1}$, write $x_1 = x_2 + s \cdot 2^{k-2}$ (why is this an improvement?). Show that for x_1 to satisfy (3) we need $s \equiv (x_2^2 - a)/2^{k-1} \pmod 2$.]

(d) Deduce that (3) has exactly 4 solutions for any $k \geq 3$.

5.10 Computing Exercise Write a program to determine all the solutions of (3) for any a (congruent to 1 mod 8) and k (within extended precision). The program should print out the successive values of s as in 5.9(c) above obtained in passing from 3 to k. When $a = 9$, why are all the values of s equal to 1? [Hint. $3^2 = 9$.] Check that your program works by testing with small values of a and k. Check that the four solutions $\pmod{2^{30}}$ of

$$x^2 \equiv 17 \pmod{2^{30}}$$

are $x = 204\,265\,751,\ 332\,605\,161,\ 741\,136\,663,\ 869\,476\,073$.

Bibliography

Those references marked by || are textbooks which are suitable for parallel (or further) reading.

H. Abelson and A. diSessa, *Turtle Geometry*, Cambridge, Mass.: MIT Press 1980 (paperback edition 1986).

J. Alanen, O. Ore and J. Stemple, 'Systematic computations on amicable numbers', *Math. Comp.* **21** (1967), 242–245.

|| R.B.J.T. Allenby and E.J. Redfern, *Introduction to Number Theory with Computing*, London: Edward Arnold 1989.

J. Anderson, 'Seeing induction at work', *Math. Gazette* **75** (1991), 406–414.

A.H. Beiler, *Recreations in the Theory of Numbers*, New York: Dover Publications, second edition 1966.

E.R. Berlekamp, J.H. Conway and R.K. Guy, *Winning Ways for your Mathematical Plays, Vol. 2: Games in Particular,* London: Academic Press 1982.

D.W. Boyd, *Amer. Math. Monthly* **97** (1990), 411–412.

R.P. Brent and G.L. Cohen, 'A new lower bound for odd perfect numbers', *Math. Comp.* **53** (1989), 431–437.

J. Brillhart and J.L. Selfridge, 'Some factorizations of $2^n \pm 1$ and related results', *Math. Comp.* **21** (1967), 87–96; corrigendum p. 751.

J.W. Bruce, 'A really trivial proof of the Lucas–Lehmer primality test', *Amer. Math. Monthly* **100** (1993), 370–371.

|| J.W. Bruce, P.J. Giblin and P.J. Rippon, *Microcomputers and Mathematics*, Cambridge, U.K.: Cambridge University Press 1990.

|| R.P. Burn, *A Pathway into Number Theory*, Cambridge, U.K.: Cambridge University Press, 1982.

G. Chrystal, *Algebra*, 7th edition New York: Chelsea Publishing Company 1964.

R.F. Churchhouse, 'Covering sets and systems of congruences', pp. 20–36 in *Computers in Mathematical Research*, Ed. R.F. Churchhouse and J.-C. Herz, Amsterdam: North-Holland 1968.

B. Cipra, 'Big number breakdown', *Science* **248** (29 June 1990), 1608.

P. Costello, 'Amicable numbers of the form $(i, 1)$', *Math. Comp.* **56** (1991), 859–865.

|| H. Davenport, *The Higher Arithmetic*, Cambridge, U.K.: Cambridge University Press, fifth edition, 1982.

L.E. Dickson, *A History of the Theory of Numbers*, New York: Chelsea Publishing Company 1966.

J. Dixon, 'The number of steps in the Euclidean algorithm', *J. Number Theory* **2** (1970), 414–422.

F. Dodd and R. Peele, 'Some counting problems involving the multinomial expansion', *Math. Magazine* **64** (1991), 115–122.

H. Dubner, 'A new method for producing large Carmichael numbers', *Math. Comp.* **53** (1989), 411–414.

H.M. Edwards, *Fermat's Last Theorem*, New York: Springer-Verlag 1977.

A. Ehrlich, 'The periods of the Fibonacci sequence modulo m', *Fibonacci Quarterly* **27** (1989), 11–13.

K.E. Eldridge and S. Sagong, 'The determination of Kaprekar convergence and loop convergence of all three-digit numbers', *Amer. Math. Monthly* **95** (1988), 105–112.

D.H. Fowler, *The Mathematics of Plato's Academy: a New Reconstruction*, Oxford: Clarendon Press 1987 (paperback 1991).

V. Gardiner, R. Lazarus, N. Metropolis and S. Ulam, 'On certain sequences of integers defined by sieves', *Math. Magazine* **29** (1956), 117–122.

M. Gardner, 'Mathematical Games' *Scientific American*, August 1970.

M. Gardner, *Penrose Tiles to Trapdoor Ciphers*, New York: W.H. Freeman 1989.

C.F. Gauss, *Disquisitiones Arithmeticae*, Translation by A.A. Clarke, revised by W.C. Waterhouse, New York: Springer-Verlag 1986.

P. Goetgheluck, 'Computing binomial coefficients', *Amer. Math. Monthly* **94** (1987), 360–365.

R. Graham, D.E. Knuth and O. Patashnik, *Concrete Mathematics*, Reading, Mass.: Addison-Wesley 1989.

A. Granville, 'Primality testing and Carmichael numbers', *Notices Amer. Math. Soc.* **39** (1992), 696–700.

H. Gupta, 'Chains of quadratic residues', *Math. Comp.* **25** (1971), 379–382.

R.K. Guy, *Unsolved Problems in Number Theory*, New York: Springer-Verlag 1981.

R.K. Guy, C.B. Lacampagne, and J.L. Selfridge, 'Primes at a glance', *Math. Comp.* **48** (1987), 183–202.

P. Hagis, Jr., 'Unitary amicable numbers', *Math. Comp.* **25** (1971), 915–918.

|| G.H. Hardy and E.M. Wright, *An Introduction to the Theory of Numbers*, Oxford: Oxford University Press, first edition 1938, fifth edition 1979.

D. Hawkins and W.E. Briggs, 'The lucky number theorem', *Math. Magazine* **31** (1957/8), 277–280.

B. Hayes, 'On the ups and downs of hailstone numbers', *Scientific American*, January 1984, 10–16.

M.E. Hellman, 'The mathematics of public-key cryptography', *Scientific American*, August 1979, 146–157.

L. Hoehn and J. Ridenhour, 'Summations involving computer-related functions', *Math. Magazine* **62** (1989), 191–196.

J.P. Jones, 'Diophantine representation of the Fibonacci numbers', *Fibonacci Quarterly* **13** (1975), 84–88.

J.H. Jordan, 'Self producing sequences of digits', *Amer. Math. Monthly* **71** (1964), 61–64.

D.R. Kaprekar, 'An interesting property of the number 6174', *Scripta Math.* **21** (1955), 304.

M. Kline, *Mathematical Thought from Ancient to Modern Times*, New York: Oxford University Press 1972.

D.E. Knuth, T. Larrabee and P.M. Roberts, *Mathematical Writing*, Math. Assoc. of America Notes No. 14, M.A.A., Washington D.C. 1989.

|| N. Koblitz, *A Course in Number Theory and Cryptography*, New York: Springer-Verlag, 1987.

M. Kraitchik, 'On the factorization of $2^n \pm 1$', *Scripta Math.* **17** (1952), 39–52.

J.C. Lagarias, 'The $3x+1$ problem and its generalizations', *Amer. Math. Monthly* **92** (1985), 3–23.

M. Lal, 'Primes of the form $n^4 + 1$', *Math. Comp.* **21** (1967), 245–247.

E.J. Lee and J.S. Madachy, 'The history and discovery of amicable numbers, parts 1, 2 and 3', *J. Recreational Math.* **5** (1972), 77–93, 153–173, 231–249.

D.H. Lehmer, 'Tests for primality by the converse of Fermat's theorem', *Bull. Amer. Math. Soc.* **33** (1927), 327–340.

H. Lehning, 'Computer-aided or analytic proof?', *College Math. J.* **21** (1990), 228–239.

M.E. Lines, *A Number for your Thoughts*, Bristol: Adam Hilger 1986.

W. McDaniel, 'Some pseudoprimes and related numbers having special forms', *Math. Comp.* **53** (1989), 407–409.

F. Morain, 'On the least common multiple of the differences of eight primes', *Math. Comp.* **52** (1989), 225–229; **54** (1990), 911.

M.A. Morrison and J. Brillhart, 'A method of factoring and the factorization of F_7', *Math. Comp.* **29** (1975), 183–205.

|| I. Niven, H.S. Zuckerman and H.L. Montgomery, *An Introduction to the Theory of Numbers*, New York: John Wiley, fifth edition 1991.

H.A. Priestley, *Introduction to Complex Analysis*, Oxford: Oxford University Press, second edition 1990.

H. Rademacher and D. Goldfeld, *Higher Mathematics from an Elementary Point of View*, Boston, Mass.: Birkhäuser 1983.

S. Ratering, 'An interesting subset of the highly composite numbers', *Math. Mag.* **64** (1991), 343–346.

P. Ribenboim, *The Book of Prime Number Records*, New York: Springer-Verlag 1988 (revised 1990).

H. Riesel, *Prime Numbers and Computer Methods for Factorization*, Boston, Mass.: Birkhäuser 1985 (revised and corrected second printing 1987).

|| K.H. Rosen, *Elementary Number Theory and its Applications*, Reading, Mass.: Addison-Wesley, second edition 1988.

M.I. Rosen, 'A proof of the Lucas–Lehmer test', *Amer. Math. Monthly* **95** (1988), 855–856.

O. Sacks, *The Man who Mistook his Wife for a Hat*, London: Duckworth 1985 (paperback edition London: Picador 1986).

A. Salomaa, *Public Key Cryptography*, New York: Springer-Verlag 1990.

I.J. Schoenberg, 'On vector indices mod m', *Math. Magazine* **61** (1988), 246–252.

M.R. Schroeder, *Number Theory in Science and Communication*, Berlin: Springer-Verlag, second edition 1986.

D. Shanks, 'A low density of primes', *J. Recreational Math.* **5** (1971), 272–275.

D. Shanks, 'Five number-theoretic algorithms', pp. 51–70 in *Proc. Second Manitoba Conference on Numerical Mathematics, Oct. 5–7, 1972*, Ed. R.S.D. Thomas and H.C. Williams, Winnipeg: Utilitas Mathematica Pub. Co.

|| D. Shanks, *Solved and Unsolved Problems in Number Theory*, New York: Chelsea Publishing Company, third edition 1985.

H.N. Shapiro, 'An arithmetic function arising from the ϕ-function', *Amer. Math. Monthly* **50** (1943), 18–30.

|| H.M. Stark, *An Introduction to Number Theory*, Cambridge, Mass.: MIT Press 1970 (paperback edition 1987).

S.M. Turner, 'Square roots mod p', *Amer. Math. Monthly*, **101** (1994), 443–449.

S. Wagon, 'The Euclidean algorithm strikes again', *Amer. Math. Monthly* **97** (1990), 125–129.

C.R. Wall, 'The fifth unitary perfect number', *Canad. Math. Bull.* **18** (1975), 115–122.

D.D. Wall, 'Fibonacci series modulo m', *Amer. Math. Monthly* **67** (1960), 525–532.

D. Wells, *The Penguin Dictionary of Curious and Interesting Numbers*, London: Penguin 1986.

W.J. Wong, 'Powers of a prime dividing binomial coefficients', *Amer. Math. Monthly* **96** (1989), 513–517.

J. Young and A. Potler, 'First occurrence prime gaps', *Math. Comp.* **52**(1989), 221–224.

Index

Note. Routine references to other textbooks are not indexed here. Some multi-author works are indexed only under the first author. See the list of references for more details. Boldface references are to definitions and theorems.

Index of Listed Programs

Index of Notation